Heftreihe des Instituts für Bauingenieurwesen
Book Series of the Department of Civil Engineering
Technische Universität Berlin

Herausgeber:
Editors:

    Prof. Dr.-Ing. Matthias Barjenbruch
    Prof. Dr.-Ing. Karsten Geißler
    Prof. Dr.-Ing. Reinhard Hinkelmann
    Prof. Dr.-Ing. Wolfgang Huhnt
    Prof. Dr.-Ing. Bernd Kochendörfer
    Prof. Dr.-Ing. Yuri Petryna
    Prof. Dr.-Ing. Stavros Savidis
    Prof. Dr. sc. techn. Mike Schlaich
    Prof. Dr.-Ing. Volker Schmid
    Prof. Dr.-Ing. Frank U. Vogdt

Shaker Verlag
Aachen 2010

Heftreihe des Instituts für Bauingenieurwesen
Book Series of the Department of Civil Engineering
Technische Universität Berlin

Band 4

Dr.-Ing. Mirko Schankat

DiaTrans – A Multi-Component Model for Density-Driven Flow, Transport and Biogeochemical Reaction Processes in the Subsurface

D 83 (Diss. TU Berlin)

Shaker Verlag
Aachen 2010

Bibliographic information published by the Deutsche Nationalbibliothek
The Deutsche Nationalbibliothek lists this publication in the Deutsche
Nationalbibliografie; detailed bibliographic data are available in the internet at
http://dnb.d-nb.de.

Zugl.: Berlin, Techn. Univ., Diss., 2009

**DiaTrans – A Multi-Component Model for Density-Driven Flow,
Transport and Biogeochemical Reaction Processes in the Subsurface**

Dissertationsschrift von Mirko Schankat
Fakultät VI – Planen, Bauen, Umwelt
der Technischen Universität Berlin

Gutachter:	Prof. Dr. Uwe Tröger
	Prof. Dr.-Ing. Reinhard Hinkelmann
	Prof. Dr.-Ing. Michael Schlüter (AWI Bremerhaven)

Tag der wissenschaftlichen Aussprache: 06.05.2009

Copyright Shaker Verlag 2010
All rights reserved. No part of this publication may be reproduced, stored in a
retrieval system, or transmitted, in any form or by any means, electronic,
mechanical, photocopying, recording or otherwise, without the prior permission
of the publishers.

Printed in Germany.

ISBN 978-3-8322-8985-0
ISSN 1868-8357

Shaker Verlag GmbH • P.O. BOX 101818 • D-52018 Aachen
Phone: 0049/2407/9596-0 • Telefax: 0049/2407/9596-9
Internet: www.shaker.de • e-mail: info@shaker.de

Abstract

The investigation of the transition zone between subsurface sediments and the water column, in addition to an enhanced understanding of the underlying processes, becomes more and more important when dealing with ecological questions not only in coastal areas.

Therefore, *DiaTrans* (**dia**genetic **trans**port), a one-phase/multi-component model to simulate multi-dimensional fully coupled density-driven flow, transport and biogeochemical reaction processes in the subsurface underlying a seawater column, including bioturbation and bioirrigation, is presented. The components include water and an arbitrary number of dissolved constituents such as chloride, methane, sulphate or oxygen. The governing equations are discretized using a *Finite-Volume-Method* (FVM) on two-dimensional rectangular structured grids in an object-oriented framework. A fully-upwinding technique is employed for the advective fluxes. The sparse and non-symmetric system of highly non-linear equations is solved, utilizing the *Newton-Raphson* method with an inner linear preconditioned BiCGSTAB solver.

This implicit fully-coupled formulation of physical and biogeochemical processes, which has certain advantages, is a new approach not only in near-shore sediments, as generally decoupled methods are employed. The model accounts for different domain sizes and grid resolutions, unsteady conditions to simulate tidal cycles and variable physical parameters such as permeabilities and porosities which are important to cope with lateral heterogeneities.

Typical benchmark tests have been carried out to verify *DiaTrans*' capability to model flow, transport and reaction processes in the subsurface. These include the *Henry* problem and the *salt dome* problem for density-driven flow and transport as well as an example for standard diagenetic biogeochemical reaction processes.

Abstract

As natural test cases for the numerical modeling different application examples are presented. First, the physical processes at *sand boils*, which account for submarine groundwater discharges at a site in the Wadden Sea of Cuxhaven, North Sea, Germany are investigated and the simulation results are compared to field measurements which have been carried out by the Alfred Wegener Institute for Polar and Marine Research (AWI), Bremerhaven, Germany. Second, reaction influenced methane concentration profiles at so-called *vent* and *partial vent* sites in Eckernförde Bay, Baltic Sea, Germany are simulated with a special focus on the effects of bioturbation and bioirrigation. For both applications, rather good matches are observed between model results and field data. The last application example deals with qualitative statements about the interaction processes at the sediment-water interface where the influence of surface water velocities on the fate of methane in the water column is investigated, considering biogeochemical reaction processes such as methane re-oxidation by oxygen and sulfate.

Because of its object-oriented nature, the model can easily be extended to additional physical processes, more complex biogeochemical reaction processes or enhanced numerics in the future. *DiaTrans* is considered as a sophisticated model for multi-dimensional density-driven flow, transport and biogeochemical reaction processes in porous media for both small and larger scale problems.

Kurzfassung

Der Schwerpunkt der Arbeit liegt in der Verbesserung des Prozessverständnisses für den Interaktionsbereich Untergrund und freier Wassersäule und darauf aufbauend in einer verbesserten Quantifizierung von Stoffflüssen, welche beispielsweise für ökologische Fragestellungen in Küstengebieten erforderlich ist.

Dazu wurde *DiaTrans* (**dia**genetic **trans**port), ein mehrdimensionales, voll gekoppeltes Ein-Phasen/Mehr-Komponenten-Modell zur Simulation von dichtegetriebenen Strömungs-, Transport- und biogeochemischen Reaktionsprozessen im Untergrund, entwickelt. Auch typische küstennahe Prozesse wie Bioturbation und Bioirrigation werden darin explizit berücksichtigt. Neben der Hauptkomponente Wasser kann eine beliebige Anzahl gelöster Substanzen, wie z. B. Chlorid, Sulfat, Methan oder Sauerstoff, betrachtet werden. Mit Hilfe eines objektorientierten Ansatzes werden die Bilanzgleichungen mit einer *Finiten-Volumen* Methode (FVM) auf zweidimensionalen strukturierten Rechteckgittern diskretisiert, wobei eine Fully-Upwind-Methode für die advektiven Flussterme verwendet wird. Das daraus resultierende unsymmetrische System aus stark nichtlinearen Gleichungen wird mit einem *Newton-Raphson* Verfahren mit innerem linearen BiCGSTAB-Gleichungslöser berechnet.

Diese implizite und vollständige Kopplung von Strömungs-, Transport- und Reaktionsprozessen stellt einen neuen Ansatz auf diesem Gebiet der Forschung dar. Das Modell berücksichtigt neben verschiedenen Gebietsgrößen und Netzauflösungen auch instationäre Bedingungen zur Simulation von Tidezyklen. Physikalische Parameter wie Durchlässigkeiten und Porositäten sind frei wählbar, um auch lokale Heterogenitäten berücksichtigen zu können.

Um die Leistungsfähigkeit von *DiaTrans* auf dem Gebiet der dichtegetriebenen Strömungs-, Transport- und Reaktionssimulation zu gewährleisten, wurde das Modell mit

Kurzfassung

typischen Benchmark-Tests verifiziert. Die vielfältigen Anwendungsmöglichkeiten des Modells werden anhand von praxisrelevanten Fragestellungen aufgezeigt.

Zum einen werden die physikalischen Prozesse an submarinen Grundwasseraustrittsstellen im Sahlenburger Watt, Nordsee, untersucht und die Simulationsergebnisse mit Feldmessungen des Alfred-Wegener-Instituts für Polar und Meeresforschung (AWI) verglichen. Zum anderen wird der Einfluss der Reaktionskinetik auf Methankonzentrationsprofile an submarinen Grundwasseraustrittsstellen in der Ostsee simuliert, wobei ein Hauptaugenmerk auf dem Einfluss der Bioturbation und Bioirrigation liegt. Für beide Anwendungsbeispiele konnten gute Übereinstimmungen zwischen Simulation und Naturmessungen erzielt werden. Darüber hinaus können mit *DiaTrans* qualitative Aussagen über die Vermischungsprozesse von gelösten Substanzen im Interaktionsbereich von Untergrund und freier Wassersäule getroffen werden. Dieser Ansatz wird anhand von Methan, mit expliziter Berücksichtigung der Oxidationsprozesse durch Sauerstoff und Sulfat, erläutert.

Durch seine objektorientierte Struktur kann *DiaTrans* in der Zukunft um weitere physikalische Prozesse, komplexere Reaktionsprozesse oder um numerische Verfahren in einfacher Weise erweitert werden. *DiaTrans* genügt hohen Ansprüchen auf dem Gebiet der Simulation von dichtegetriebenen Strömungs-, Transport und Reaktionsprozessen in Sedimenten und porösen Medien sowohl auf kleinen als auch auf größeren räumlichen Skalen.

Acknowledgements

This dissertation summarizes three years of my work in the field of numerical simulation in subsurface flow at the Chair of Water Resources Management and Modeling of Hydrosystems, Department of Civil Engineering, Technische Universität Berlin and at the Section of Marine Geochemistry, Alfred Wegener Institute for Polar and Marine Research, Bremerhaven.

I would like to thank most specially Professor Reinhard Hinkelmann and Professor Michael Schlüter for giving me the opportunity to work on this project, their permanent support, great freedom in carrying out my work, and, finally, for acting as main advisors.

The discussions with my colleagues in Berlin contributed to a big extent to the progress of this work. Their helpfulness, qualifications and collaboration as well as the long-standing good working atmosphere will always remain in my mind.

I specially want to thank my colleague Leopold Stadler for his excellent suggestions, help and support, particularly at the beginning of my work in Berlin.

I also want to thank Vikramjeet Singh Notay and Andy-David Jablonski, who were my Master's and Diploma students and who contributed well to this work.

Also many thanks to my parents Brigitte and Günter for their support throughout my academic career.

Finally, I especially thank my wife Kathleen for her understanding and support, particularly in the last few months, when I was occupied too much with finalizing my thesis. Kathleen - you made me a better person and you still do.

Berlin, 2009 - Mirko Schankat

Contents

Abstract i

Kurzfassung iii

Acknowledgements v

Contents vii

Nomenclature xi

List of Figures xix

List of Tables xxv

1 Introduction 1
   1.1 Motivation 1
      1.1.1 Submarine Groundwater Discharge - An Interesting Phenomenon . . 1
      1.1.2 Deficits – The Need for a New Numerical Model 5
   1.2 Objectives & Organization of this Work 8

2 Study Area & Measurements 11
   2.1 Study Area Description 11
      2.1.1 Location and Characteristics 11
      2.1.2 The Freshwater Lense 13
   2.2 Measurement Techniques 14
      2.2.1 Pore Water Sampling with Suction Candles 15
      2.2.2 Pore Water Sampling with Rhizons 15

Contents

	2.2.3 Submarine Groundwater Discharge Measurements	17
3	**Model Concepts, Physics & Biogeochemistry in Near-Shore Sediments** . .	**19**
3.1	The Continuum Approach for Porous Media in the Subsurface	20
3.2	Fluid & Soil Properties	22
	3.2.1 Fluid Phase and Components	22
	3.2.2 Component Fraction	22
	3.2.3 Density	23
	3.2.4 Viscosity	24
	3.2.5 Porosity	24
	3.2.6 Permeability	25
3.3	Overview of Physical Processes	27
3.4	Advection	29
3.5	Diffusion & Dispersion	30
	3.5.1 Diffusion	31
	3.5.2 Dispersion	32
3.6	Overview of Biogeochemical Processes	36
3.7	Bioturbation	38
3.8	Bioirrigation	39
3.9	Reactions	41
	3.9.1 Governing Reactions	42
	3.9.2 Governing Component Reaction Balance Equations	46
	3.9.3 Reaction Kinetics	48
3.10	The Multi-Component Formulation	53
	3.10.1 Continuity Equation	53
	3.10.2 Momentum Equation	56
	3.10.3 Density Dependency	57
4	**Numerical Methods & Implementation**	**61**
4.1	Object-Oriented Programming in Java	61
	4.1.1 Advantages	62
	4.1.2 Disadvantages	63
	4.1.3 Implementation in DiaTrans	64
4.2	Finite-Volume-Method	67

4.3	Discretization of Rate of Accumulation	69
	4.3.1	One-Step Methods	70
	4.3.2	Implementation	71
4.4	Discretization of Advection	72
	4.4.1	Implementation	72
	4.4.2	Upwinding Techniques	73
4.5	Discretization of Diffusion & Dispersion	75
4.6	Discretization of Bioturbation	76
4.7	Discretization of Bioirrigation	77
4.8	Discretization of Reactions	78
4.9	Discretization of Sinks & Sources	78
4.10	Initial and Boundary Conditions	79
	4.10.1	Initial Conditions	79
	4.10.2	Boundary Conditions	79
4.11	Solving Nonlinear Systems of Equations	81
	4.11.1	Introduction	81
	4.11.2	Newton-Raphson Method	82
	4.11.3	BiCGSTAB Method	84
	4.11.4	Preconditioning	86
4.12	Adaptive Time-Stepping	87
	4.12.1	Introduction	87
	4.12.2	Implementation	88
4.13	Pre- & Postprocessing	90
	4.13.1	Preprocessing	90
	4.13.2	Postprocessing	90

5 Verification ... 95

5.1	The Henry Problem	95
	5.1.1	Introduction	95
	5.1.2	Model Setup	96
	5.1.3	Results	98
5.2	The Salt Dome Problem	101
	5.2.1	Introduction	101
	5.2.2	Model Setup	102
	5.2.3	Results	103

	5.3	Standard Reaction Processes 106
	5.3.1	Introduction 106
	5.3.2	Model Setup 108
	5.3.3	Results 110

6 Applications 115

    6.1 Modeling Physical Processes at Sand Boils 116

        6.1.1 General Remarks on MUFTE_UG 116

        6.1.2 Sensitivity Analysis 118

        6.1.3 Simulation vs. Nature 122

        6.1.4 Tidal Influence 125

        6.1.5 DiaTrans vs. MUFTE_UG 134

    6.2 Modeling Reaction Processes at SGD Sites 137

        6.2.1 Introduction 137

        6.2.2 Effects of Bioirrigation in Advective Flow 139

        6.2.3 Model Setup Eckernförde Bay 144

        6.2.4 Results 147

    6.3 Modeling Interactions with the Water Column 155

        6.3.1 Introduction 155

        6.3.2 Approach & Model Setup 156

        6.3.3 Results 161

7 Summary, Conclusions & Outlook 167

Bibliography 173

Nomenclature

abbreviations

1D	one-dimensional
2D	two-dimensional
BC	boundary condition
BiCGSTAB	Biconjugate Gradient Stabilized Method
BLAS	Basic Linear Algebra Subprograms
eq.	equation
fig.	figure
FVM	Finite-Volume-Method
IC	initial condition
ILU	incomolete LU factorization
LAPACK	Linear Algebra PACKage
OM	organic matter
REV	representative elementary volume
sec.	section

Nomenclature

SGD submarine groundwater discharge

exponents

c component

CH_4 methane

Cl chloride

Fe^{2+} dissolved iron

H_2O pure water

H_2S dissolved sulfide

Mn^{2+} dissolved manganese

n current time level

$n+1$ new time level

NO_3 nitrate

O_2 oxygen

SO_4 sulfate

C center cell of control volume

E east neighbor cell of control volume

e east interface of control volume

N north neighbor cell of control volume

n north interface of control volume

S south neighbor cell of control volume

s south interface of control volume

W	west neighbor cell of control volume	
w	west interface of control volume	

terms with Greek letters

α_{bi1}	constant for bioirrigation term	$[1/s]$
α_{bi2}	constant for bioirrigation term	$[1/m]$
α_l	longitudinal dispersion length	$[m]$
α_t	transversal dispersion length	$[m]$
Δx	width of cell in horizontal x-direction	$[m]$
Δy	width of cell in vertical y-direction	$[m]$
Δz	depth of cell in third coordinate direction	$[m]$
Γ	boundary of domain Ω	
μ	dynamic viscosity	$[kg/(ms)]$
Ω	solution domain	
ϕ	porosity	$[-]$
ϕ_s	solid porosity	$[-]$
ρ_{mass}^f	pure freshwater mass density	$[kg/m3]$
ρ_{mass}^s	pure saltwater mass density	$[kg/m3]$
ρ_{mass}	mass density	$[kg/m^3]$
ρ_{mol}	molar density	$[mol/m^3]$
τ	turtuosity	$[-]$
θ	Crank-Nicholson factor	$[-]$

Nomenclature

ε infinitesimal small value for Newton-Raphson method ....

terms with Latin letters

$\underline{\underline{D}}_{bt}$ bioturbation tensor .... $[m^2/s]$

$\underline{\underline{D}}_d$ mechanical dispersion tensor .... $[m^2/s]$

$\underline{\underline{D}}_{hyd}$ hydrodynamic dispersion tensor .... $[m^2/s]$

$\underline{\underline{D}}_{m,e}$ effective molecular diffusion tensor .... $[m^2/s]$

$\underline{\underline{K}}$ intrinsic permeability .... $[m^2]$

$\underline{\underline{k}}$ effective permeability .... $[m^2]$

$\underline{\underline{K}}_f$ hydraulic conductivity .... $[m/s]$

$\underline{F}$ flux term ....

$\underline{F}$ vector of functions ....

$\underline{g}$ gravitational vector .... $[m/s^2]$

$\underline{J}_a$ advection flux .... $[mol/(sm^2)]$

$\underline{J}_{bt}$ bioturbation flux .... $[mol/(sm^2)]$

$\underline{J}_d$ dispersion/diffusion flux .... $[mol/(sm^2)]$

$\underline{n}$ normal vector .... $[-]$

$\underline{v}$ Darcy velocity .... $[m/s]$

$\underline{v}$ velocity .... $[m/s]$

$\underline{v}_a$ seepage velocity .... $[m/s]$

$\underline{w}$ diffusive flux ....

$\underline{x}$ local vector .... $[m]$

$\underline{x}$	vector of unknowns	
A	system matrix	
B	vector of the right-hand side	
b	depth in the sediment column	$[m]$
B_i	bioirrigation term	$[mol/(sm^3)]$
C	matrix	
C^c_{mass}	mass concentration	$[kg/m^3]$
C^c_{mol}	molar concentration	$[mol/m^3]$
Cl	chlorinity	$[‰]$
D_{bt}	bioturbation coefficient	$[m^2/s]$
$D_{m,e}$	effective molecular diffusion coefficient	$[m^2/s]$
D_m	molecular diffusion coefficient	$[m^2/s]$
e	scalar entity	$[-]$
$E(t)$	extensive state variable	$[-]$
F_i	functional relation	
$fact1$	factor for adaptive time-stepping	$[-]$
$fact2$	factor for adaptive time-stepping	$[-]$
G	conventional symbol for organic matter concentration	$[mol/m^3]$
g	gravitational accelaration	$[m/s^2]$
I	concentration of inhibiting constituent	$[mol/m^3]$
k	rate constant for organic matter decomposition	$[1/s]$

Nomenclature

K_i'	inhibition constants	$[mol/m^3]$ or $[-]$
K_i	Monod saturation constants	$[mol/m^3]$
k_j	reaction rate constants	$[m^3/(mol\,s)]$
k_r	relative permeability	$[-]$
L	lower matrix	
l	length scale of REV	$[m]$
M^c	molecular weight	$[kg/mol]$
O	surface area of boundary Γ	
p	pressure	$[Pa]$
$p, \overline{p}$	vectors	
q^c	source / sink term	$[mol/(s m^3)]$
r	sink or source	
$r, \overline{r}$	vectors	
r^c	reaction term	$[mol/(s m^3)]$
R_i	reaction rate term	$[mol/(m^3 s)]$
S	salinity	$[‰]$
$s, \overline{s}$	vectors	
T	temperature	$[°C]$
t	time	$[s]$
t	vector	
U	upper matrix	
V	volume	$[m^3]$

v	vector	
v_{ax}	seepage velocity in x-direction	$[m/s]$
v_{ay}	seepage velocity in y-direction	$[m/s]$
X	mass fraction	$[-]$
X	vector of unknowns	
x	mole fraction	$[-]$
x, y, z	coordinates in the global coordinate system	$[m]$
x_0	porewater mole fraction	$[-]$
x_i	unknown values	
$x_{t=0}^c$	initial condition for component c	$[-]$

List of Figures

1.1	Schematic overview of submarine groundwater discharge (after BURNETT et al., 2003)	2
2.1	Overview of the study area Wadden Sea of Cuxhaven, North Sea, Germany (left after LINKE, 1979; right after LGN, 2000)	12
2.2	Closer look at the measurement area (after KURTZ, 2004)	12
2.3	Results of aeroelecromagnetic measurements with detailed view of the freshwater lense (after SIEMON and BINOT, 2001)	14
2.4	Left: Blank suction candle produced by UMS Munich; Right: Suction candle in use in the sediment (both after SCHARF, 2008)	15
2.5	Left: Rhizons with filter areas and extraction needle, (after SCHARF, 2008); Right: Peristaltic pump (after KURTZ, 2004)	16
2.6	Left: Rhizon pole, (after SCHARF, 2008); Right: Rhizon grid to sample on a cross-sectional profile (after KURTZ, 2004)	16
2.7	Discharge measurement chamber with plastic bag and beaker (after SCHARF, 2008)	17
3.1	Definition of the *REV* model concept (after BEAR, 1972 and HELMIG, 1997)	21
3.2	Left: Sand boils in the Wadden Sea of Cuxhaven from top (after SCHARF, 2008); Right: Sand boil from bottom (after KURTZ, 2004)	28
3.3	Advection / convection, diffusion and dispersion processes (after BARLAG et al., 1998)	29
3.4	Actual path length (solid line) and smallest distance between two points (dashed line) in a porous medium (after CIRPKA, 2005)	31

List of Figures

3.5	Reasons for variability of the flow field and dispersion on different spatial scales (after KINZELBACH, 1992)	33
3.6	Scale dependency of longitudinal dispersion length α_l (after KINZELBACH, 1992)	34
3.7	Overview of biogeochemical and physical processes in the sediment (modified after HAESE, 1999)	36
3.8	Function of bioturbation coefficient vs. depth (based on formulation after FOSSING et al., 2004)	39
3.9	Function of bioirrigation term vs. depth (based on formulation after SCHLÜTER et al., 2000)	40
3.10	Qualitative distribution of solid concentrations (OM, MnO_2 and $Fe(OH)_3$)	52
3.11	Control volume Ω (after HELMIG, 1997)	54
4.1	Simplified flow chart of *DiaTrans*	65
4.2	Control volumes of the FVM (modified after HINKELMANN, 2005)	67
4.3	Compass notation for rectangular structured two-dimensional grid	69
4.4	Principles of solving nonlinear systems of equations (modified after CLASS, 2004)	82
4.5	Pseudocode of the BICGSTAB Method (modified after LEPEINTRE, 1992)	86
4.6	Input-File-Editor of *DiaTrans*	91
4.7	Exemplary output results of *Paraview* for the Henry problem	92
5.1	Model setup *Henry* problem	97
5.2	Chloride concentration distributions during the Henry problem	98
5.3	*Darcy* velocity distribution of the *Henry* problem	99
5.4	Result comparison *Henry* problem (modified after OSWALD et al., 1996)	100
5.5	Model setup *salt dome* problem	102
5.6	Chloride concentration distribution for the *salt dome* problem at steady-state	104
5.7	*Darcy* velocity distribution and streamlines for the *salt dome* problem	104
5.8	Result comparison for the *salt dome* problem (solutions from HERBERT et al., 1988; *FEFLOW* and *ROCKFLOW* (both after KOLDITZ et al., 1998) taken from DIERSCH and KOLDITZ, 2005)	105

List of Figures

5.9   Typical concentration profiles for standard reaction processes (modified after JOURABCHI et al. (2005) – top left; BOUDREAU (1996) – top right & WANG and VAN CAPPELLEN (1996) – bottom left and right) ... 107
5.10 Model setup for standard reaction processes in columns 108
5.11 Solid concentration profiles as used for verification of standard reaction processes (expressed per unit volume total sediment.) 110
5.12 Concentration distribution for standard reaction processes (without bioturbation & bioirrigation) 111
5.13 *DiaTrans* concentration profiles for standard reaction processes without (top) and with bioturbation & bioirrigation (bottom) 112
6.1   Model setup and boundary conditions (after SCHANKAT et al., 2007) 119
6.2   Chloride concentration profiles / advection (left) – diffusion/dispersion (middle) – no SGD influence (right) (after SCHANKAT et al., 2007) .. 120
6.3   Left: Chloride concentration profiles at $x = 1.00\,m$ for different freshwater inflow widths; Right: Different molecular diffusion coefficients (both after SCHANKAT et al., 2007) 122
6.4   Seepage velocity distribution and vectors (after SCHANKAT et al., 2007) 124
6.5   Water density distribution (after SCHANKAT et al., 2007) 124
6.6   Chloride concentration profiles (after SCHANKAT et al., 2007) 125
6.7   Model setup (after SCHANKAT et al., 2009b) 126
6.8   Chloride concentration distribution and velocity vectors during low tide (after SCHANKAT et al., 2008a) 128
6.9   Chloride concentration distribution and velocity vectors during high tide (after SCHANKAT et al., 2008a) 129
6.10 Freshwater flux (Su_flux) & saltwater height vs. time [MUFTE_AWI_20_06] (after SCHANKAT et al., 2009b) 130
6.11 Estimation of corresponding saltwater height to zero freshwater flux (after SCHANKAT et al., 2009b) 131
6.12 Freshwater flux (Su_flux), freshwater mass (mass_Su) & saltwater mass (mass_Sa) vs. time [MUFTE_AWI_20_06] (after SCHANKAT et al., 2009b) 131

xxi

List of Figures

6.13 Chloride concentration profiles over depth at x= 1.0 m for different pressures at lower boundary and varying saltwater height (0.0 – 0.8 m) [MUFTE_AWI_20_04-06] (after SCHANKAT et al., 2009b) 132

6.14 Maximum freshwater fluxes (after SCHANKAT et al., 2009b) 133

6.15 Chloride concentration profile comparison $MUFTE\_UG$ and $DiaTrans$ at $x_4 = 1.00\,m$ 136

6.16 Location and bathymetry of a vent (after SAUTER, 2001) 138

6.17 Model setup for effects of bioirrigation 140

6.18 Methane and sulfate concentration distributions for simulation A without bioirrigation (top) and for simulation B with bioirrigation (bottom) 143

6.19 Model setup Eckernförde Bay 145

6.20 Top: Pressure distribution; Bottom: Chloride concentration distribution and seepage velocity vectors (both for simulation C at a $vent$ location) 147

6.21 Top: Methane concentration distribution; Bottom: Sulfate concentration distribution (both for simulation C at a $vent$ location) . 148

6.22 Top: Methane concentration distribution; Bottom: Sulfate concentration distribution (both for simulation D at a $partial\ vent$ location) 149

6.23 Comparison of simulation results (C and D) and measurements for methane and sulfate concentration profiles 151

6.24 Typical methane concentration profiles in Eckernförde Bay at locations highly influenced by SGD (left) and at locations less influenced by advective flow (right) (after SCHLÜTER et al., 2004) 152

6.25 Comparison of simulation results (D and E) and measurements for methane concentration profiles 154

6.26 Model setup interaction processes 157

6.27 $Darcy$ (flow) velocity distribution (Scenario B) 162

6.28 Concentration distributions for methane and a conservative tracer (Scenario A) 163

6.29 Concentration distributions for methane and a conservative tracer (Scenario B) 164

6.30 Concentration distributions for methane and a conservative tracer (Scenario C) 165

List of Tables

3.1 Typical values for porosity (after FREEZE and CHERRY, 1979) 25
3.2 Typical values for effective permeability (after FREEZE and CHERRY, 1979) 26
3.3 Biogeochemical parameters and rate constants for coastal zones 51

5.1 Physical parameters for the *Henry* problem 97
5.2 Physical parameters for the *salt dome* problem 103
5.3 Physical parameters for the standard reaction processes 109

6.1 Boundary conditions for the various simulation setups (after SCHANKAT et al., 2009*b*) 127
6.2 Physical parameters for effects of bioirrigation 142
6.3 Physical parameters for the simulation at a *vent* (simulation C) and at a *partial vent* (simulation D) 146
6.4 Parameters for scenarios A, B and C for the simulation of interaction processes 161

1
Introduction

The chapter starts with an overview of the overall motivation of this research, including the explanation of the importance of the submarine groundwater discharge phenomenon and information about the need for a new numerical model to simulate the governing processes related to flow, transport and reactions at submarine groundwater discharge sites in near-shore sediments or in porous media in general. The second part outlines the main objectives as well as the structure of this work.

1.1 Motivation

Submarine groundwater discharge is an interesting phenomenon which can be observed all around the world. The understanding of the governing processes including the physical processes such as flow and transport as well as biogeochemical reaction processes is of major importance when dealing with ecological questions in near-shore sediments. Utilizing numerical simulations to understand and predict the physical and biogeochemical processes occurring in the subsurface, including the fate of constituents, is also becoming more and more important when dealing with ecological questions in this field.

1.1.1 Submarine Groundwater Discharge - An Interesting Phenomenon

The definition of submarine groundwater discharge (SGD) is not unique, different common descriptions exist next to each other. Following an oceanographic view, all water

1 Introduction

fluxes below the sediment surface is seen as groundwater, including recirculated seawater. TANIGUCHI et al. (2002) describe SGD as "(...) all direct discharge of subsurface fluids across the land-ocean interface". A hydrological interpretation of SGD according to JOHANNES (1980) can be formulated as "(...) all freshwater discharge below the high tide mark, including water discharge from the beach above sea level at low tide". As from this hydrological standpoint, groundwater is only seen as water from an aquifer, this definition only includes the groundwater itself, neglecting the recirculated seawater. A schematic overview of SGD including the processes leading to pore water exchange is shown in fig. 1.1.

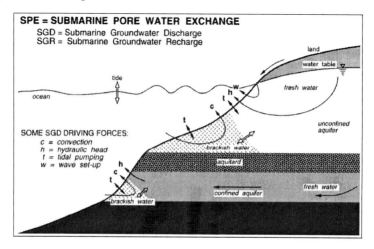

Figure 1.1: Schematic overview of submarine groundwater discharge (after BURNETT et al., 2003)

Although the phenomenon of SGD is an often undetected problem which is also hard to quantify because of its small discharge rates, it has been discovered by field measurements in several coastal zones all around the world, in the Baltic Sea, the North Sea, the Mediterranean Sea, the Gulf of Mexico, the Sea of Japan or the Yellow Sea, for example. The significance of SGD for hydrological budgets as well as the transport and release of contaminants, nutrients or trace gases was underlined by several studies (JOHANNES, 1980; VALIELA et al., 1990; LAROCHE et al., 1997). Also a number of researchers investigated the phenomenon, including the detection and quantification using direct (e. g. seepage meters) and indirect (e. g. isotope or conductivity) measurement techniques (MOORE, 1996; BURNETT et al., 2003; TANIGUCHI et al., 2006).

Generally, SGD can occur everywhere where an aquifer is hydraulically connected with the seawater. It is required that the piezometric head inside the aquifer is higher than the water surface of the sea so that a hydraulic gradient is induced which reaches seawards and that the sediment is permeable enough to let the freshwater pass through. According to JOHANNES (1980), SGD often occurs at shelf areas, where permeable glacial sediments can be found. If the submarine groundwater discharge is arising on a small spacial scale in very permeable sediments, the flow velocities of SGD are much higher than the flow velocities inside the aquifer. BOKUNIEWICZ (1980) and JOHANNES (1980) state that 40 - 98% of SGD occurs in a range of about 100 m from the coast line for unconfined aquifers. When talking about confined aquifers, the extent can be much larger and SGD will occur when the overlying aquitard is locally disturbed and therefore more permeable.

SGD is influenced by different factors. First of all, terrestrial impacts which govern the flow dynamics and the location of the groundwater table are important. This can be precipitation, evaporation and discharge rates as they affect the groundwater recharge and therefore the hydraulic gradient (SCHARF, 2008). Parameters such as porosity and permeability of the subsurface sediments also influence the rate of SGD as they control flow velocity and direction. All of the above mentioned control the rate of SGD at the sediment-water interface.

On the other hand, tidal influence and wind velocities are of major importance as they impact wave conditions near the shore. As BURNETT et al. (2003) state, stratification of the water body as well as density differences also control the rate of SGD. The differences between mostly advective transport through *sand boils* or diffusive / dispersive transport of SGD is explained in more detail in sec. 3.3.

SCHARF (2008) has done an admirable job of collecting information about the importance of different aspects of the phenomenon, underlining that SGD is interesting for current and further research. Some of those aspects are briefly summarized in the following paragraphs.

One is, that dissolved constituents can be enriched in groundwater compared to the overlying seawater. Submarine groundwater discharge can transport those constituents and nutrients very fast from the sediment to the bottom water across the sediment-water interface. This flux of dissolved organic compounds and nutrients is very important for the productivity and the mineralization processes in the sediment and at the sediment surface (SCHARF, 2008). According to RIEDL and MACHAN (1972) and BUSSMANN et al. (1999), it is also determining the micro-climatic conditions for the benthic micro and macro fauna. Furthermore, the nutrient exchange is enhanced in the sediment within the different sedimentological and geochemical areas (SCHLÜTER et al., 2004).

Also methane and carbon dioxide can be released into the water column and atmosphere. Pockmarks, which are also associated with SGD have been investigated in the Eckernförder Bay, Baltic Sea, Germany (BUSSMANN et al., 1999). According to SCHLÜTER (2002) pockmarks can be formed due to gas-related eruptions, submarine groundwater discharge or anthropogenic interference and are morphological crater-like structures in the sediment. WHITICAR (2002) associated those pockmarks with periodically occurring SGD in the Eckernförder Bay. BUSSMANN et al. (1999) detected enriched methane concentrations in those areas due to higher microbial activity, whereas SCHLÜTER et al. (2004) did not recognize an explicit interrelation between SGD and methane release.

Nitrate concentrations in SGD which are two to three times higher than typical concentrations in near-shore seawater were obtained at Perth, Western Australia. This leads to higher productivity of benthic algae, which can only be explained with SGD (JOHANNES, 1980). LAPOINTE et al. (1997) also found out that nitrate enriched groundwater influences the growth of giant algae to the disadvantage of coral reefs at the coastal zones of Jamaica and Florida. BREIER (2006) states that in Jamaica giant algae are already dominating the ecosystem instead of corals.

According to CHARETTE (2004), dissolved nitrogen concentrations can be up to 100 to 1000 times higher in groundwater than in seawater. With this mostly anthropogenic influence, SGD can have ecological effects, although the discharge might be small compared to the one from surface waters. SGD can have an especially large impact in coastal areas, where the circulation of the water body is inhibited, e. g. in almost closed bays or inland waters. HWANG et al. (2005) conducted tracer studies in Bangdu Bay, Korea and identified SGD as the main source of eutrophication.

Finally, the effects of SGD on benthos population due to the lower salinity concentrations is not negligible. ZIPPERLE and REISE (2005) carried out extensive studies at List, Sylt, North Sea, Germany and showed that the reduced salinity in the pore waters caused by SGD has influenced the population density of the lug (*Arenicola marina*).

Instead of the adult lug, the estuary ragworm (*Nireis diversicolor*) and the king rag (*Nireis virens*) as well as the juvenile lug have settled (ZIPPERLE and REISE, 2005). Therefore, *Arenicola marina* could be used as an indicator for SGD at this site.

1.1.2 Deficits – The Need for a New Numerical Model

As the investigation of the transition zone between subsurface sediments and the seawater column becomes more and more important when dealing with ecological questions not only in coastal areas, the numerical modeling of flow, transport and reaction processes in the sediment and across the sediment-water interface is one of the main objectives in this field of research. Other objectives are the quantification of the nutrient recycling, of competing bacterial processes affecting the decomposition of organic matter and of the admixture of chemical constituents into the seawater column with special focus on the water quality.

Another important task is the improved estimation of solute fluxes in highly heterogeneous areas. Several diagenetic 1D-models such as CANDI (BOUDREAU, 1996) or STEADYSED1 (WANG and VAN CAPPELLEN, 1996) are very suitable for detailed consideration of reaction processes, whereas the transport of fluids and dissolved constituents by coupled advective and diffusive / dispersive processes, as they occur especially in areas affected by *sand boils*, is often implemented in a rather simplified manner. Most of the diagenetic models in use so far only deal with a simplified flow field, i.e. the multi-dimensional character of flow and transport cannot be reflected due to the one-dimensional nature of those simulating tools. Therefore, the combined advective and diffusive / dispersive processes can not be simulated as they occur in nature at SGD sites.

The herein presented model *DiaTrans* (**dia**genetic **trans**port) utilizes a novel approach in this context. Because of its two-dimensional one-phase/multi-component structure it is possible to model flow, transport and biogeochemical reaction processes fully coupled in more detail and more realistic in the vicinity of SGD sites as well as generally for small and bigger scale problems.

Additionally, density-dependent flow is incorporated in the model, which makes it even more suitable for modeling processes occurring in near-shore sediments, especially during tidal cycles. Other important biogeochemical processes occurring in sediments, such as bioturbation and bioirrigation, are also included. The model is not restricted to the number of components and diagenetic reactions, i.e. as many dissolved constituents as necessary, such as chloride, methane, sulfate or oxygen, can be regarded and new reactions can be added easily.

Although well developed numerical models for flow and transport and reaction processes in the subsurface already exist, e.g. *MODFLOW* (HARBAUGH, 2005), *FEFLOW* (DIERSCH, 1995), *ROCKFLOW* (KOLDITZ et al., 1999) or *MT3D* (ZHENG, 1990), the new model *DiaTrans* has certain advantages in this field. As mentioned before, additional important biogeochemical processes such as bioturbation and bioirrigation are incorporated in the code in contrast to the other models. Also, a fully coupled multi-component formulation for all relevant physical and biogeochemical processes is employed in *DiaTrans* compared to the above mentioned models, which utilize decoupled approaches. The benefits of the fully-coupled formulation are outlined in the following.

Flow, transport and reaction processes in sediments can be simulated in different ways. If all processes are coupled within one system of discretized equations, it is called a *direct coupling* approach. It is also possible to solve for the flow field first, which is used afterwards to solve the transport equation for advection and diffusion / dispersion. These results can then be employed to solve for the reaction processes. Note, that for all processes different time-steps are used. Such a system is called decoupled and an *iterative two-step* method (see CIRPKA, 1997) can be utilized for example. The *operator-splitting* technique is another technique which is often employed to solve for flow and transport processes in porous media. In this approach, the basic equations are split into different parts, e.g. into an advective and diffusive part (see HINKELMANN, 2005 for further information).

In general, the *direct coupling* approach yields advantages compared to the decoupled methods, i.e. *operator-splitting* technique and *iterative two-step* method. CIRPKA (1997) gives a good overview of the main advantages and disadvantages of those three methods. According to CIRPKA (1997), the *operator-splitting* approach induces a numerical error due to the time-lagged coupling. It is stated that the main disadvantage compared to the *direct coupling* approach is that all of the simulated processes occur at the same time in nature and interact with each other, but are modeled decoupled as a sequence of processes. The numerical error can not be estimated and this decoupling can lead to instability during the solving procedure. The prime downside of the *iterative two-step* technique is that convergence can not be guaranteed a priori for any chosen time step size, as the choice of the time step size is constrained by the coupling scheme (CIRPKA, 1997). Therefore, the choice of time-step sizes is crucial in these two methods, compared to the *direct coupling* approach in which only one time-step size is employed for all processes.

The *direct coupling* technique has another major advantage compared to the two methods mentioned before. An always consistent linearization of the whole problem of partial differential equations can be achieved at all times. Although computation time can be much higher using *direct coupling* due to the very large systems of equations, *DiaTrans* uses this technique as it produces the most accurate, stable and thus realistic and reliable results. The disadvantage of higher computational effort can partly be overcome by using adaptive time-stepping techniques as also utilized in *DiaTrans*.

Besides the fully coupled formulation of physical and biogeochemical processes mentioned above, there is another main advantage of *DiaTrans* compared to other numerical codes. Usually, it is fairly complicated to extend existing models or the possibilities to fit the model to the needs of certain applications is limited. Because of its object-oriented modular structure, it is fairly easy in *DiaTrans* to incorporate even more physical processes or additional biogeochemical reactions, which have not been regarded by now. Any reaction which can be described with one of the two main reaction types (*Monod-type* and *second-order* kinetics) can easily be implemented directly (see sec. 3.9). Additionally, other reaction kinetics approaches than the ones mentioned, can also be implemented in a rather simple manner. This makes *DiaTrans* very suitable to cover different fields of application dealing with flow, transport and reaction processes not only in near-shore sediments, but generally in porous media.

1 Introduction

1.2 Objectives & Organization of this Work

The main objective of this work is the development of a new numerical model to simulate flow, transport and reaction processes in sediments. The model *DiaTrans* shall be able to overcome certain deficits of current models, not only for the modeling in near-shore sediments but in porous media in general. *DiaTrans* includes a multi-dimensional fully coupled one-phase / multi-component approach, including density effects on flow and transport processes. Current models for the simulation of biogeochemical reaction processes are often of one-dimensional nature and only employ a simplified flow field. As especially at submarine groundwater influenced sites advective processes cannot be neglected, the multi-dimensional multi-component formulation utilized in *DiaTrans* is new in this field of research. With this approach, better estimates for the quantification of dissolved component fluxes in subsurface sediments are possible.

Although typical processes only occurring in coastal environments such as bioturbation and bioirrigation are included, the model can generally be applied to a wide field of applications. The simulation of flow, transport and reaction processes is not limited to coastal environments as *DiaTrans* can be utilized for subsurface flow, transport and reaction problems in porous media in general, also on larger scales.

Another objective of this work is the application of *DiaTrans* to simulate typical issues related to near-shore sediments. This includes the modeling of physical processes and biogeochemical reaction processes at submarine groundwater discharge influenced sites. The obtained simulation results presented herein shall be compared to existing field measurements to verify the reliability of the model not only in this special field of research.

The work is divided into seven chapters:

- Chapter 1 presents an introduction, including the motivation of this work. The phenomenon of submarine groundwater discharge is explained in detail, as well as the deficits of current numerical models in the field of subsurface flow, transport and reaction modeling. The resulting objectives for this work follow.

- Chapter 2 introduces the study area in the Wadden Sea of Cuxhaven, North Sea, Germany. Also information about measurement techniques to obtain porewater concentration profiles or submarine groundwater discharge rates is given.

1.2 Objectives & Organization of this Work

- In chapter 3, the underlying model concepts of the newly developed numerical model *DiaTrans* are explained. Therefore, information about the background of the physical and biogeochemical processes occurring in subsurface sediments is given. The chapter closes with an explanation about the multi-component formulation employed in this work.

- The numerical methods utilized in *DiaTrans* as well as their implementation is outlined in chapter 4. First, the advantages and disadvantages of the programming language *Java* are depicted, followed by details about the *Finite-Volume-Method*. The discretization techniques and their implementation are summarized and information about techniques to solve highly nonlinear systems of equations and about adaptive time-stepping is given.

- Verification examples for flow, transport and reaction processes in the subsurface are presented in chapter 5. This includes the *Henry* problem, the *salt dome* problem and standard reaction processes in columns. *DiaTrans*' performance in those verification examples is discussed in detail.

- Different application examples are outlined in chapter 6. Physical processes at so-called *sand boil* sites are simulated, including the tidal effect on porewater concentration profiles. Also, biogeochemical reaction processes at submarine groundwater discharge sites are investigated, with special focus on methane concentration distribution at so-called *vent* and *partial vent* sites. This chapter closes with a rather simplified approach to investigate the physical and biogeochemical processes across the sediment-water interface.

- The contents of this work are summarized in chapter 7. Further conclusions are drawn and an outlook on possible future work is presented.

2

Study Area & Measurements

This chapter describes the study area and its characteristics, followed by a brief description of the measurement techniques for pore water concentration and submarine groundwater discharge measurements, which are used throughout the field measurements carried out by the Alfred Wegener Institute for Polar and Marine Research, Bremerhaven, Germany.

2.1 Study Area Description

In this section the study area in the Wadden Sea of Cuxhaven, North Sea, Germany is described. Information about the location and the geomorphological, geological and hydrogeological characteristics are given, as well as some background about the freshwater lense, which is the driving force for submarine groundwater discharge. More detailed description is given in KURTZ (2004) and SCHARF (2008).

2.1.1 Location and Characteristics

The study area, which is close to the village Sahlenburg, is located about 8 km west of Cuxhaven in the North Sea in Germany (see fig. 2.1, left). The Wadden Sea of Cuxhaven belongs to the National Park "Niedersächsisches Wattenmeer". The coastal area west of Sahlenburg is the eastern border of the mouth of the river Weser. East of Cuxhaven, the river Elbe flows into the North Sea (SCHARF, 2008).

The area, where most of the *sand boils* can be found, is shown in more detail in fig. 2.1 (right) and has an extent of about 3800 m times 380 m. The arrow in fig. 2.1 (right, lower part) indicates the location where most of the measurements have taken place.

2 Study Area & Measurements

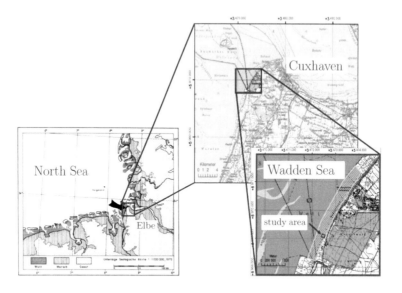

Figure 2.1: Overview of the study area Wadden Sea of Cuxhaven, North Sea, Germany (left after LINKE, 1979; right after LGN, 2000)

This area is located between two groins and has an extent of about 105 m times 140 m. Fig. 2.2 depicts a closer look at this smaller scale measurement area.

Figure 2.2: Closer look at the measurement area (after KURTZ, 2004)

Sand boils, often also referred to as *seeps*, are "crater"-like structures through which SGD can enter the seawater column. Although each of them is small in size, they discharge a large amount of freshwater across the sediment-water interface, as a high number of *sand boils* can be found even in such a small area. A more detailed explanation of the characteristics of *sand boils* is given in sec. 3.3.

The Wadden Sea is an amphibian landscape which is present all along the coast of the North Sea. It is influenced by low and high tide and gets flooded twice a day. The sediments in the Wadden areas consist mostly of fine sand, peat and clay. Especially in the Wadden Sea of Cuxhaven, a number of peat layers can be found below the sediment surface. Organic matter inside the sediments results from rest products of plants and animals, but can also be transported to the areas from the open North Sea and the mouths of bigger rivers. Shallow areas are also wetted at low tide, as they are crisscrossed with tidal creeks.

2.1.2 The Freshwater Lense

In the Wadden Sea of Cuxhaven a freshwater lense has been found lying below the saltwater body during an "aerophysical investigation of the coastal aquifers between the mouths of river Weser and river Elbe" (RODEMANN et al., 2005) using a helicopter measuring device. During this campaign an area of saltwater intrusion and the freshwater lense could be clearly identified.

Electromagnetic measurements use the specific resistance as an indicator for electric conductivity, with "the conductivity being the reciprocal value of the specific resistance" (DE STADELHOFEN, 1995). Using this specific resistance, the salinity of the aquifer can be estimated, where saltwater has a high value of conductivity and therefore a low value of resistance and freshwater can be categorized the opposite way.

The results of the aeroelecromagnetic measurements in the Wadden Sea (*Watt*) of Cuxhaven are shown in fig. 2.3. In this figure the coast line is displayed (*Küstenlinie*) as well as areas where saltwater intrusion into the aquifer takes place (*Salzwasserintrusion*) and the location of the freshwater lense (*Süßwasserlinse*), where the study area is located. The reddish colored areas (as at the area of saltwater intrusion) indicate low values of specific resistance and therefore higher values of salinity, whereas the greenish to bluish colored areas indicate high values of specific resistance with low values of salinity (as at the freshwater lense and at the mainland). For more detailed information about the aeroelecromagnetic measurements and the technique refer to SIEMON and BINOT (2001), KURTZ (2004) and SCHARF (2008).

The freshwater lense indicates that groundwater is present below the seawater column along the coastal area outlined in fig. 2.3. As this reservoir is of confined nature, freshwater is able to enter the seawater column across the sediment-water interface through *sand boils* due to the upward hydraulic gradient. The SGD rates are higher at

2 Study Area & Measurements

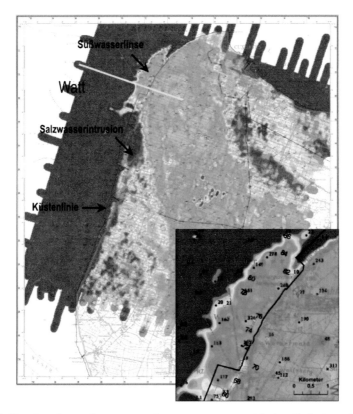

Figure 2.3: Results of aeroelecromagnetic measurements with detailed view of the freshwater lense (after SIEMON and BINOT, 2001)

times of low tide, and run dry during high tide as the accumulating saltwater decreases the hydraulic gradient (see also sec. 3.3).

2.2 Measurement Techniques

This section explains the measurement and sampling techniques which were used by the Alfred Wegener Institute for Polar and Marine Research to obtain a vast amount of data which is utilized for the applications in sec. 6. This includes two different pore water sampling techniques with suction candles and rhizons as well as the submarine groundwater discharge measurements. Pore water sampling was carried out to determine concentration profiles for the distribution of several constituents such as chloride,

sulfate and methane. The submarine groundwater discharge measurements were used to correlate the concentration profiles to different discharges as well as to changes in concentration during tidal cycles at *sand boils*. A detailed description of the measurement and sampling techniques described herein can be obtained from SCHARF (2008).

2.2.1 Pore Water Sampling with Suction Candles

To determine the chloride, sulfate or methane concentration of near-shore bottom water, suction candles produced by the company UMS Munich, Germany were used (fig. 2.4, left). To produce the needed vacuum, small extraction needles are used at one end of the suction candle (fig. 2.4, right). The candles can be utilized to extract bottom water up to a depth of 25 cm into the sediment with the amount of extracted water being around 4 mL. To overcome the problem of unintentional admixture or contamination of the water sample, the first 4 mL were discarded. To correlate the samples to the location in the Wadden Sea, the GPS coordinates of each sample were also saved.

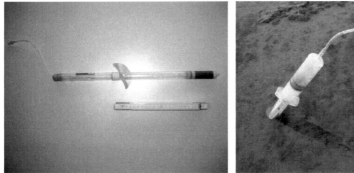

Figure 2.4: Left: Blank suction candle produced by UMS Munich; Right: Suction candle in use in the sediment (both after SCHARF, 2008)

2.2.2 Pore Water Sampling with Rhizons

For pore water sampling, a newly developed Rhizon in situ sampler (RISS) (SEEBERG-ELVERFELDT et al., 2005) was used. Rhizons have been utilized in soil sciences for a long time to measure the soil moisture content. These rhizons are distributed by e.g. Eijkelkamp Agrisearch Equipment and are made of hydrophilic porous polymer tubes, which are strengthened with a metal or nylon wire and then extended with a polyvinyl chloride tube (fig. 2.5, left). The polymer tube has a typical pore diameter of 0.1 to 0.2 μm. To collect the pore water through rhizons, spring-loaded syringes, vacuum tubes or peristaltic pumps can be used (fig. 2.5, right).

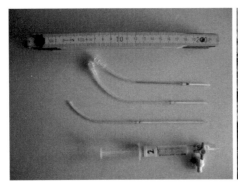

Figure 2.5: Left: Rhizons with filter areas and extraction needle, (after SCHARF, 2008); Right: Peristaltic pump (after KURTZ, 2004)

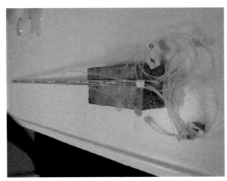
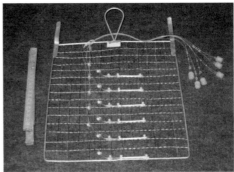

Figure 2.6: Left: Rhizon pole, (after SCHARF, 2008); Right: Rhizon grid to sample on a cross-sectional profile (after KURTZ, 2004)

Rhizons have several advantages compared to other sampling devices, e.g. low mechanical disturbance of the sediment due to the small diameter (2.4 mm), low dead volume (0.5 mL) or minimized sorption processes on the inert polymer. All major components and nutrients, such as chloride, sulfate, methane and nitrate can be analyzed. Rhizons were also employed in this research to sample constituent concentrations over depth utilizing rhizon poles (fig. 2.6, left) or on a cross-sectional profile along a *sand boil* using rhizon grids (fig. 2.6, right).

2.2.3 Submarine Groundwater Discharge Measurements

The submarine groundwater discharge measurements at *sand boils* were carried out with a simple discharge chamber (fig. 2.7) based on the principles of a Lee-type seepage meter (LEE, 1977). A plastic bag was connected to the measurement chamber through a tube connector and the amount of outflowing freshwater was measured afterwards using a simple beaker. The submarine groundwater discharge was averaged over four measurements to overcome the uncertainties of the measurement technique.

Figure 2.7: Discharge measurement chamber with plastic bag and beaker (after SCHARF, 2008)

This simple method based on the Lee-type seepage meter was chosen, as other discharge measurements techniques such as turbine sensors or photo meters did not yield reliable results (compare to DIETRICH, 2006). The sensitivity of the sensor to contamination with sediment particles and the limited measurement range are problematic for turbine sensoring, whereas data obtained from photo meters was found to be unreliable due to technical problems.

3

Model Concepts, Physics & Biogeochemistry in Near-Shore Sediments

The underlying model concepts of *DiaTrans* are outlined in this chapter. The well known continuum approach for porous media in the subsurface, which makes use of the representative elementary volume (REV) concept, is explained at the beginning, followed by the description of fluid and soil properties such as permeability, porosity or density.

Then, an overview of the main physical processes which occur in near-shore sediments is presented, including aspects such as advection and diffusion / dispersion, the consideration of density effects of the liquid phase and the influence of the tidal cycle. Especially their effect on flow and transport processes at *sand boils* is highlighted. Detailed explanations of the physical processes which are implemented in *DiaTrans* and their background follow.

The overview of the biogeochemical processes occurring at *sand boils* and in near-shore sediments in general is presented next. Especially the diagenetic reaction processes considered in *DiaTrans*, which are described using *Monod*-type and second-order kinetics, are explained. Although the character of bioturbation and bioirrigation could be thought of being of physical nature, these processes are also illustrated as biogeochemical reaction processes, as they are induced by biological inhabitants of the sediment and as they particularly influence the biogeochemistry. Also detailed explanations for these processes are given.

The chapter closes with information about the developed multi-component formulation for flow, transport and reaction processes in the subsurface. This includes the continuity equation, the momentum equation and the explanation of the density dependency.

3.1 The Continuum Approach for Porous Media in the Subsurface

In BEAR (1972), porous media are defined as a portion of space occupied by a number of phases, with at least one being a solid. The phases are distributed throughout the entire space, and a Representative Elementary Volume (REV) can be found. Sand, peat, gravel are good examples for porous media, with the main feature being that both solid matrix and pore spaces exist. The pore spaces are connected and therefore allow flow and transport through the porous medium (JABIR, 2005).

In this work the subsurface sediment system underlying a seawater column in the Wadden area is the porous medium of main interest. It mainly consists of fine sand and peat layers. Generally, *DiaTrans* is capable of modeling any subsurface system consisting of one incompressible solid and one incompressible fluid phase. The governing parameters to describe the porous system such as porosity (sec. 3.2.5) and permeability (sec. 3.2.6) can be freely chosen and distributed in the domain during the modeling process. The presented framework could also be easily extended to unsteady conditions for the porosity for example, to account for its dependence on the pressure.

State variables are used to describe the system's properties. Those state variables can be divided into *extensive* and *intensive* properties. *Extensive* properties, whose magnitudes are additive for subsystems, are e.g. mass, momentum and energy, whereas *intensive* properties, whose magnitudes are independent of the extent of the system, are variables such as temperature or pressure (JABIR, 2005).

Although porous media are highly heterogeneous having complex pore structures, it is common to simulate flow, transport and reaction processes with a volume averaged approach which treats the porous media as a "macroscopically uniform continuum" (JABIR, 2005). This implies that the "processes occurring on the micro-scale between single pores are upscaled and averaged over certain volumes which are called representative elementary volumes or $REVs$" (HINKELMANN, 2005). The length scale of

3.1 The Continuum Approach for Porous Media in the Subsurface

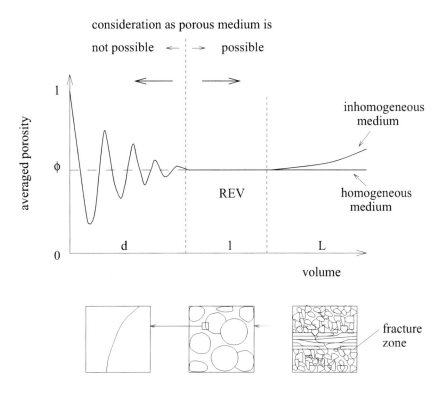

Figure 3.1: Definition of the *REV* model concept (after BEAR, 1972 and HELMIG, 1997)

an *REV l* (see fig. 3.1) is chosen, so that "it leads to a representative average of the property under consideration, e.g. the porosity ϕ" (HINKELMANN, 2005). To avoid deviations of the property, the REV must be large enough, but also small enough to account for spatial variations. According to HINKELMANN (2005), "the REV idea is the most common model concept" used for porous media and subsurface system considerations and detailed information is given in BEAR (1972) or HELMIG (1997). HINKELMANN (2005) describes, that the equivalent model concept is "(...) based on the *REV* idea and the assumption that the inhomogeneous domain can be homogenized bit by bit by shifting the observation scale". A definition of the *REV* model concept after BEAR (1972) and HELMIG (1997) is given in fig. 3.1.

3.2 Fluid & Soil Properties

In the following, some general remarks about the definitions for fluid phase and components as well as the fluid and soil properties density, viscosity, porosity and permeability utilized in this work are given. Strictly speaking, all parameters described in sec. 3.2.3 to 3.2.5 are functions of the pressure p, the temperature T or others. However, in near-surface sediments where pressure and temperature variations are small, these effects are usually neglected, as it is done in this work (compare to JABIR, 2005).

3.2.1 Fluid Phase and Components

DiaTrans is developed to simulate flow, transport and reaction processes in subsurface systems consisting of the solid (matrix) phase and a liquid/fluid phase (normally water). Although such a porous medium system consists of the two phases solid and liquid/fluid, it is often referred to as a single-phase system as only the flow, transport and reaction processes of the liquid phase are simulated.

According to JABIR (2005), "a phase may be defined as a portion of space occupied by a material whose properties are uniform (...) and is separated from other (...) materials by a well defined interface." A fluid phase usually consists of a number of components c.

A component, described by the superscript c, is a distinct chemical species, which can be either a chemical element (e.g. chloride), a molecular substance (e.g. oxygen) or a mixture of different substances (methane for example). Note, that pure freshwater (H_2O) is also defined as the main component inside the water phase. *DiaTrans* is capable of modeling any number of dissolved components inside the water phase depending on the task being performed.

3.2.2 Component Fraction

Mole fraction x or *mass fraction* X is used to describe the amount of a component c in the liquid/fluid phase. *MUFTE_UG* (e.g. HELMIG, 1997; HELMIG et al., 1998; HINKELMANN, 2005), which was employed for the preliminary studies such as the sensitivity analysis (sec. 6.1.2), utilizes the mass fraction formulation. As in *DiaTrans* also reaction processes are included, the mole fraction formulation is used, as the reaction equations are formulated stoichiometrical, i. e. in number of moles of a component. According to JABIR (2005), the definitions for mass fraction X and mole fraction x read as follows:

$$X^c = \frac{mass\,of\,c\,in\,fluid\,phase}{total\,mass\,of\,fluid\,phase} \quad [-] \tag{3.1}$$

$$x^c = \frac{number\,of\,moles\,of\,c\,in\,fluid\,phase}{total\,number\,of\,moles\,in\,fluid\,phase} \quad [-] \tag{3.2}$$

From mass conservation it follows that:

$$\sum_c X^c = 1 \tag{3.3}$$

$$\sum_c x^c = 1 \tag{3.4}$$

The molecular weight M^c, which is defined as the mass of one mole of component c, is used to characterize each component and can be employed to convert mole fractions to mass fractions and vice versa (compare to STEPHAN and MAYINGER, 1990).

JABIR (2005) describes that the "amount of a component that is present in a (fluid) phase can also be described by mass concentration $C^c_{mass}\,[kg/m^3]$ or molar concentration $C^c_{mol}\,[mol/m^3]$". Those concentration definitions read as follows:

$$C^c_{mass} = X^c \rho_{mass} \tag{3.5}$$

$$C^c_{mol} = x^c \rho_{mol} \tag{3.6}$$

where ρ_{mass} is the mass density and ρ_{mol} is the molar density (see sec. 3.2.3).

3.2.3 Density

The density of a fluid can either be described by using the *mass density* ρ_{mass} or the *molar density* ρ_{mol}. Only the dependency of the density on the salinity, i. e. the chloride concentration respectively, is regarded in *DiaTrans*, although the fluid density can generally be a function of other parameters such as pressure or temperature. Functional relationships for the density of liquids are given in INTERNATIONAL FORMULATION COMMITTEE (1967) and REID et al. (1987).

3 Model Concepts, Physics & Biogeochemistry in Near-Shore Sediments

To convert the mass density into a molar density in a fluid phase, the following relationship can be utilized:

$$\rho_{mol} = \frac{\rho_{mass}}{\sum_{c}(M^c x^c)} \qquad (3.7)$$

3.2.4 Viscosity

Viscosity is a property of a fluid phase which is being deformed by either shear stress or extensional stress. It represents the fluid's resistance to flow under the influence of the forces causing the flow (compare to JABIR, 2005). Generally, in any flow, layers move at different velocities and the fluid's viscosity arises from the shear stress between these layers. *Newtonian* fluids, such as water or gas, satisfy the assumption that the shear stress between the layers is proportional to the velocity gradient in the direction perpendicular to the layers.

The *dynamic viscosity* μ of a fluid phase, in this work further only referred to as *viscosity*, determines the dynamics of an incompressible *Newtonian* fluid and is the property of the fluid phase, whereas the *kinematic viscosity* is the *dynamic viscosity* divided by the density of a *Newtonian* fluid.

Although the *dynamic viscosity* is a function of the fluid phase composition in nature, e.g. temperature or dissolved constituents (see INTERNATIONAL FORMULATION COMMITTEE, 1967 and REID et al., 1987), it is assumed to be constant in this work ($\mu = 1.0 \cdot 10^{-3} Pa\,s$). The *dynamic viscosity* is employed to convert the *hydraulic conductivity* $\underline{\underline{K}}_f$ to the *effective permeability* $\underline{\underline{k}}$ (sec. 3.2.6) when using the *Darcy's Law*.

3.2.5 Porosity

The portion of a porous medium which is capable of storing a fluid is called the *fluid porosity* ϕ, further only referred to as *porosity*. It can be expressed as the fraction of the pore space volume to the bulk volume of the medium:

$$\phi = \frac{pore\,space\,volume}{bulk\,volume\,of\,medium} \quad [-] \qquad (3.8)$$

The porosity can further be divided into *effective* and *ineffective* porosity, as only pores which are interconnected can contribute to the fluid flow (*effective* porosity). Also, water which is bound to clay particles for example, does not contribute to the *effective porosity*. As only the effective part of the *porosity* is of interest for flow in porous media, hereafter *porosity* refers to the *effective porosity*.

In this work, deformation processes of the solid matrix, which would lead to changes in *porosity* due to pressure variations, are not taken into account. The porosities are therefore assumed to be constant over time, but different values can be assigned to different regions in the domain in *DiaTrans*.

Similar to the (fluid) porosity a *solid porosity* ϕ_s can be defined:

$$\phi_s = \frac{volume\ occupied\ by\ solids}{bulk\ volume\ of\ medium} \quad [-] \qquad (3.9)$$

In a single fluid phase system consisting of a solid matrix and one fluid phase, following definition holds:

$$\phi + \phi_s = 1 \qquad (3.10)$$

Typical values for the *porosity* ϕ are given in e.g. FREEZE and CHERRY (1979) with some of them being listed in table 3.1. Note, that generally the *effective porosity* is a lot smaller than the values given in table 3.1, especially in silty and clayey sediments (see LEGE et al., 1996 for further information).

Table 3.1: Typical values for porosity (after FREEZE and CHERRY, 1979)

Soil Material	Porosity range [-]
Gravel	0.25 - 0.40
Sand	0.25 - 0.50
Silt	0.35 - 0.50
Clay	0.40 - 0.70

3.2.6 Permeability

The *effective permeability* $\underline{\underline{k}}$ is defined as the product of the *relative permeability* k_r and the *intrinsic permeability* $\underline{\underline{K}}$ (HINKELMANN, 2005).

$$\underline{\underline{k}} = k_r \underline{\underline{K}} \qquad (3.11)$$

"The intrinsic permeability is a porous material property and characterizes the ease with which fluids can pass through porous media" (JABIR, 2005), whereas the relative permeability k_r is a dimensionless measure of the effective permeability of a fluid phase and an important factor in multi-phase flow. In a porous medium which is fully saturated with a single fluid phase, as discussed in this work, the relative permeability k_r is equal to 1.

The effective permeability $\underline{\underline{k}}$, further only referred to as permeability, is transferred to the hydraulic conductivity $\underline{\underline{K}}_f$, which depends on both the medium and fluid properties, in the following manner using the definition of eq. 3.11 with k_r being equal to 1:

$$\underline{\underline{K}}_f = \underline{\underline{k}} \frac{\rho_{mass}\, g}{\mu} \qquad (3.12)$$

Typical values for permeability $\underline{\underline{k}}$ are given in e.g. FREEZE and CHERRY (1979) with some of them being listed in table 3.2.

Table 3.2: Typical values for effective permeability (after FREEZE and CHERRY, 1979)

Soil Material	Permeability range [m^2]
Gravel	$10^{-7} - 10^{-10}$
Clean Sand	$10^{-9} - 10^{-12}$
Silty Sand	$10^{-10} - 10^{-14}$

3.3 Overview of Physical Processes

When talking about physical flow and transport processes in sediments, one can distinguish between mostly advective, mostly diffusive / dispersive processes or a combination of both. Advective flow and transport can be forced by pressure differences but also by density differences in the liquid phase. Generally, advection is of minor importance in near-shore sediments. Pressure differences are low, as the height of the saltwater column is small and large density differences do not occur as the sediment is mostly fully saturated either with saltwater or freshwater. When regarding tidal effects with their bigger changes in absolute pressure, advection can have a bigger influence and can even be a major factor around SGD sites where large density differences occur due to submarine freshwater with possibly high discharge rates.

Compared to advection, the effects of diffusion and dispersion are generally much higher. Although these processes are slower and the discharges are smaller, the overall contribution can be quite high, as diffusion and dispersion is also present in areas not directly influenced by SGD and therefore in the whole sediment. It is not forced by pressure differences but by concentration differences of constituents, which are always present in the sediment column.

The important effect of density-induced flow forced by density differences in the liquid phase is also implemented in *DiaTrans*, which is a major difference to other flow and transport models. This makes *DiaTrans* very suitable to model the processes in near-shore sediments, especially when talking about the simulation of flow and transport at SGD sites and the freshwater-seawater-interactions in their near surroundings. Density differences can be either a result of temperature differences but also because of differences in concentrations, e.g. concentration of chloride. In *DiaTrans* only the latter effect is implemented, as the temperature distribution in the sediment is not explicitly regarded.

The influence of tidal changes, which is very important especially at the near shore, can be considered in *DiaTrans* by using time-dependent boundary conditions. It is very easy in *DiaTrans* to set up a sinusoidal type of boundary condition for the water pressure for example. This increases the field of application of the model even more, as time dependent flow problems can also be simulated and analyzed in detail.

One of the key phenomena regarded in this work are the processes at SGD sites, especially around *sand boils*. As mentioned before, submarine groundwater discharge often occurs through so-called *sand boils*, e. g. at the study area in the Wadden Sea of Cuxhaven, but also at different sites in the Baltic Sea. These *sand boils* can be described as small freshwater (groundwater) "springs" in the uppermost part of coastal sediments with discharges up to 2 L/min. It is assumed that at most sites a confined freshwater aquifer reaches seawards and lies beneath the seawater along the coast. At places with higher permeable areas, favoured flow paths are formed due to the pressure gradient and *sand boils* up to 20 to 50 cm in diameter form up (fig. 3.2, left). Sometimes fine sediment is washed out due to the highly advective freshwater flux as can be seen in fig. 3.2 (right). Although the granular structure is fluidized in nature in these cases, *DiaTrans* assumes the soil always to be of consolidated nature.

 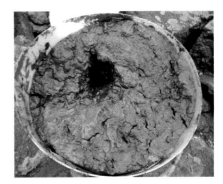

Figure 3.2: Left: Sand boils in the Wadden Sea of Cuxhaven from top (after SCHARF, 2008); Right: Sand boil from bottom (after KURTZ, 2004)

As the whole system is driven by differences in water density and big pressure gradients due to the confined aquifer, advection plays a more important role than compared to regular near-shore sediments. Advection controls the processes mainly in the *sand boil* center, whereas diffusion and dispersion processes are more important away from the *sand boils*. A combination of those processes is also always present in the near surroundings of those structures. The numerical model must be capable of simulating both processes very detailed and reliable, which *DiaTrans* surely can, due to its fully coupled multi-component formulation which takes into account that freshwater and saltwater are fully miscible. This is an important feature, as the width of the transition zone between regions of different chloride concentrations and therefore densities cannot be neglected in such systems (e. g. HINKELMANN et al., 2000a).

The background of the main physical processes advection and diffusion / dispersion is explained in detail in the following (see also HINKELMANN and HELMIG, 2002).

3.4 Advection

One of the main processes of solute transport in subsurface flow is advection (convection), which is "the transport of particles at the same speed and into the same direction as the average groundwater [fluid] flow" (FESEKER, 2004), either in "horizontal or vertical direction without changing the shape of the concentration isoareas" (HINKELMANN, 2005) (see fig. 3.3, upper part). Consequently, the flux due to advection is a function of the groundwater or fluid flow inside a fluid phase.

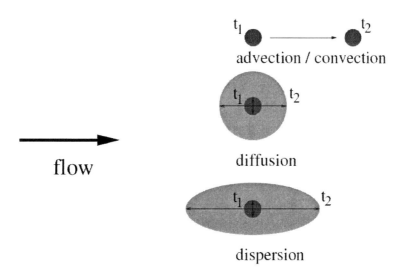

Figure 3.3: Advection / convection, diffusion and dispersion processes (after BARLAG et al., 1998)

In *DiaTrans*, advection can be either due to pressure gradients or density differences. Again, with the latter implementation, which is presented in sec. 3.10.3, it is accounted for density-driven flow and transport which is especially important at SGD sites in near-shore sediments.

The advective flux across an interface of a certain control volume can be described as:

$$\underline{J}_a = x^c \rho_{mol} \underline{v} \tag{3.13}$$

The flow velocity of a phase is obtained by replacing the momentum equation with the generalized *Darcy* law (see sec. 3.10.2). Substituting eq. 3.64 into eq. 3.13 leads to eq. 3.14:

$$\underline{J}_a = -x^c \rho_{mol} \frac{\underline{\underline{k}}}{\mu} (grad\, p - \rho_{mass} \underline{g}) \tag{3.14}$$

The numerical implementation and discretization of eq. 3.14 in the *Finite-Volume-Method (FVM)* is explained in detail in sec. 4.4.

3.5 Diffusion & Dispersion

Besides advection, *diffusion* and *dispersion* are two other main transport processes in porous media. Their effects can be considered using the hydrodynamic dispersion tensor $\underline{\underline{D}}_{hyd}$, which is the sum of the effective molecular diffusion tensor $\underline{\underline{D}}_{m,e}$ and the mechanical dispersion tensor $\underline{\underline{D}}_d$ (see eq. 3.15). The diffusion / dispersion term, which describes the diffusive / dispersive flux, is characterized in *DiaTrans* as shown in eq. 3.16. Both processes are based on the *First Fickian* law. As diffusive / dispersive processes of dissolved constituents can only take place in the pore space, the flux due to diffusion and dispersion is multiplied with the porosity (see eq. 3.16):

$$\underline{\underline{D}}_{hyd} = \underline{\underline{D}}_{m,e} + \underline{\underline{D}}_d \tag{3.15}$$

$$\underline{J}_d = -\phi \rho_{mol} \underline{\underline{D}}_{hyd}\, grad\, x^c \tag{3.16}$$

In the following, the main features of diffusion and dispersion processes are outlined. A detailed description of the numerical implementation and discretization of this term in the FVM is presented in sec. 4.5.

3.5.1 Diffusion

Diffusion processes are a result of *Brown*'s molecular movement which leads to a compensation of concentration (here: mole fraction) differences. Therefore, this purely physical transport process is always in the direction of lower concentrations (see HINKELMANN, 2005). As can be seen from fig. 3.3 (middle part), diffusive spreading of a constituents is independent of the direction.

According to CIRPKA (2005), in porous media the molecular diffusion coefficient D_m is divided by the turtuosity τ, which describes the ratio of the actual path length of a particle from one point to another to the smallest distance of those two points (see fig 3.4).

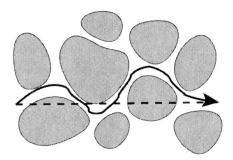

Figure 3.4: Actual path length (solid line) and smallest distance between two points (dashed line) in a porous medium (after CIRPKA, 2005)

This leads to the definition of the effective molecular diffusion coefficient (eq. 3.17) which is used inside the effective molecular diffusion tensor (eq. 3.18). For simplicity reasons, $\tau = 1$ in *DiaTrans*, which leads to $D_{m,e} = D_m$.

$$D_{m,e} = \frac{D_m}{\tau} \tag{3.17}$$

The effective molecular diffusion tensor $\underline{\underline{D}}_{m,e}$ has only entries of the effective molecular diffusion coefficient on its main diagonal with the other terms being 0 (eq. 3.18):

$$\underline{\underline{D}}_{m,e} = \begin{pmatrix} D_{m,e} & 0 \\ 0 & D_{m,e} \end{pmatrix} \tag{3.18}$$

An average value for the molecular diffusion coefficient D_m for dissolved constituents in porous media is $\approx 1.0 \cdot 10^{-9}\, m^2/s$ (e. g. LEGE et al., 1996). In *DiaTrans* it is possible to define different molecular diffusion coefficients for each component which are constant for the whole computational domain.

3.5.2 Dispersion

HINKELMANN (2005) states that "dispersion represents all transport effects which are caused by inhomogeneities of the flow field below the REV scale". The effect of dispersion is greater in the direction of flow than transversal to it (see fig. 3.3, lower part).

According to HINKELMANN (2005), "for bigger REVs, fluctuating velocities due to inhomogeneities of the aquifer [or porous medium in general] lead to macrodispersion" (see fig. 3.5, middle and right). CIRPKA (2005), HINKELMANN (2005) and KINZELBACH (1992) give reasons for the different aspects which cause these variabilities of the flow field on the microscale (fig. 3.5, left).

First of all, dispersion is a result of the laminar pipe flow like parabolic velocity profile within a pore which causes *Taylor-Aris* dispersion (fig. 3.5, left – upper part). It has to be mentioned that this phenomenon is only in the scale of molecular diffusion or even smaller.

The second effect which causes dispersion is due to the fact that the pore sizes are variable in size. Compared to the *Taylor-Aris* dispersion, this leads to the fact that the dispersive transport along some streamlines can be much faster than along others (fig. 3.5, left – middle part).

Due to the flow around the grains itself, the flow and therefore the transport of the constituents is deflected transversely (fig. 3.5, left – lower part) which also leads to a dispersive flux (SAFFMAN, 1959, 1960; DE JOSSELIN DE JONG, 1958).

The mechanical dispersion tensor $\underline{\underline{D}}_d$ is anisotropic even in an isotropic medium (HINKELMANN, 2005) and can be described in a 2D case as in eq. 3.19. This formulation after SCHEIDEGGER (1961) is the most common description of the mechanical dispersion tensor with its entries being outlined in eq. 3.20:

3.5 Diffusion & Dispersion

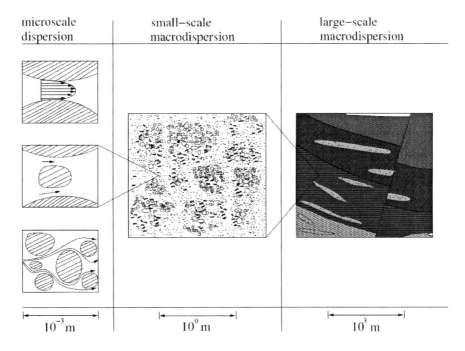

Figure 3.5: Reasons for variability of the flow field and dispersion on different spatial scales (after KINZELBACH, 1992)

$$\underline{\underline{D}}_d = \begin{pmatrix} D_{xx} & D_{xy} \\ D_{yx} & D_{yy} \end{pmatrix} \tag{3.19}$$

$$\begin{aligned} D_{xx} &= \alpha_l \frac{v_{ax}^2}{|\underline{v}_a^2|} + \alpha_t \frac{v_{ay}^2}{|\underline{v}_a^2|} \\ D_{xy} &= D_{yx} = (\alpha_l - \alpha_t) \frac{v_{ax} v_{ay}}{|v_a|} \\ D_{yy} &= \alpha_t \frac{v_{ax}^2}{|\underline{v}_a^2|} + \alpha_l \frac{v_{ay}^2}{|\underline{v}_a^2|} \end{aligned} \tag{3.20}$$

In eq. 3.20, α_l and α_t stand for the longitudinal and transversal dispersion lengths. Generally, the dispersion in flow direction (i.e. α_l) is about one order of magnitude larger than dispersion in transversal direction (i.e. α_t). $\underline{v}_a$ denotes the seepage velocity,

v_{ax} the seepage velocity component in horizontal x-direction across an interface and v_{ay} the component of the seepage velocity in vertical y-direction across an interface, all of them being obtained using the *Darcy* law (eq. 3.64 and eq. 3.65). From the definition above it is clear, that the entries of the hydrodynamic dispersion tensor (D_{xx} for example) are calculated directly at an interface.

Because of the reasons for variability described above, it is obvious that dispersion is highly scale-dependent. According to HINKELMANN (2005), the longitudinal dispersion length was found to be $0.0001 < \alpha_l < 0.01\,m$ for homogeneous sands on the laboratory scale, whereas it ranges from $0.07 < \alpha_l < 0.7\,m$ for natural subsurface systems, e.g. gravel. As a consequence of macrodispersion (fig. 3.5, middle and right), which results from inhomogeneities of the aquifer, the dispersion lengths can be about 4 to 5 orders of magnitude higher on the field scale. Also on the field scale, molecular diffusion is often negligible compared to mechanical dispersion (HINKELMANN, 2005).

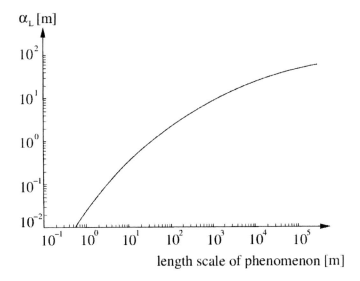

Figure 3.6: Scale dependency of longitudinal dispersion length α_l (after KINZELBACH, 1992)

As macrodispersion increases with the length of the phenomenon, i.e. the transport way, fig. 3.6 can be used to estimate the longitudinal dispersion length. According to HINKELMANN (2005), "(...) this effect is limited as soon as the inhomogeneities are represented by the computation of the flow field". Therefore, an upper limit for the dispersion lengths exist.

Note that for practical applications, the dispersion lengths are calibration parameters which contain all other unknown conditions, e.g. missing information about geological structures or effects which are not included in the model concept. For further information about dispersion see KINZELBACH (1992).

3.6 Overview of Biogeochemical Processes

The biogeochemical processes which occur in subsurface sediments can be influenced by advective and diffusive / dispersive processes described in sec. 3.3. As *DiaTrans* uses a fully-implicit and therefore fully coupled multi-component formulation, it is capable of simulating all these inter-dependencies directly, which is an important difference to other models dealing with flow and transport processes in porous media. Standard reactions such as sulfate reduction, methane decomposition and aggregation, denitrification or oxygen consumption, which are highly coupled to each other, but also bioturbation and bioirrigation are thought of as processes affecting the biogeochemistry. All these aspects just mentioned are implemented in *DiaTrans*.

An overview of all aspects governing the biogeochemistry of constituents which are either solved in water (dissolved) or solid constituents (particulate) in the subsurface is given in fig. 3.7. The main physical aspects such as advection and diffusion / dispersion are also outlined.

In fig. 3.7, the redox boundary (**red**uction-**ox**idation boundary) denotes, that iron or manganese is present as oxidized species ($Fe(III)$ or $Mn(IV)$) above this boundary whereas below the reduced species occur ($Fe(II)$ or $Mn(II)$). Although only the reaction aspects of manganese and iron are shown in this figure after HAESE (1999), these processes could be replaced with any other reaction mentioned above.

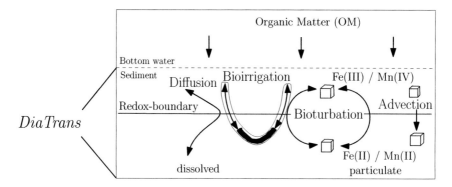

Figure 3.7: Overview of biogeochemical and physical processes in the sediment (modified after HAESE, 1999)

The degradation of organic matter (OM), which accumulates in the uppermost layers of the sediment column, is described and implemented by the so called "Primary Reactions" (see also sec. 3.9.1). According to FOSSING et al. (2004), the order of these five reactions reflects "(...) the energy obtained by the bacteria by degrading organic matter using oxygen, nitrate, manganese, iron and sulphate". Therefore, the degradation process by oxygen is the respiration process which yields most of the energy, whereas the process of sulfate respiration yields least of the energy. For the primary redox reactions, *Monod*-type kinetics are usually adopted (BOUDREAU, 1996, 1997; BOUDREAU and WESTRICH, 1984) as it is also done in *DiaTrans*.

Secondary reactions for the dissolved constituents including the suggested kinetic rate laws are regarded using information collected by VAN CAPPELLEN and WANG (1995, 1996). These reactions can be described as "second order chemical reactions" with two species reacting irreversibly at a rate k. The *Monod*-type kinetics as well as the kinetics for the secondary reactions are implemented in a general form, so that it is easily possible to regard as many reactions for as many constituents as necessary for different fields of application.

Bioturbation induces fluxes which result from mixing in a biological way of thinking, i.e. it is produced by active organisms living in the sediment. In reality, processes such as tube excavation, ingestion and excretion of sediment lead to a movement of both fluid and sediment, which again then influences the biogeochemistry in the sediment. Bioturbation is actually a phenomena of discrete and local events. But as we normally look at an area of study which has a larger extent, bioturbation can be described and is implemented in *DiaTrans* using a formulation, which is similar to the one of diffusion / dispersion processes.

Another important feature occurring in nature and included in *DiaTrans* is bioirrigation. In the uppermost fauna-inhabited sediment layers this phenomenon can be observed. Irrigating animals, which live in small tubes or small channels, play an important role in the quick and easy exchange of solutes between sediment and bottom water through these small canals and also affect the biogeochemical reaction processes due to this exchange of constituents in the liquid phase. Compared to bioturbation and diffusion the transport via bioirrigation is much faster, as the solutes do not have to pass all sediment layers as they take a shortcut through the channels and tubes (compare to FOSSING et al., 2004).

The background of the main biogeochemical processes bioturbation, bioirrigation and reactions is explained in detail in the following.

3.7 Bioturbation

Bioturbation, sometimes also called biodiffusion, induces fluxes which result from mixing in a biological way of thinking, i. e. it is produced by active organisms living in the sediment. In reality, those processes such as tube excavation, ingestion and excretion of sediment, lead to a movement of both fluid and sediment. As we are only interested in this modeling approach in the mixing of the fluid, we neglect the processes acting on the solids.

As mentioned before, bioturbation is actually a phenomenon of discrete and local events. But as we normally look at an study area of larger extent, bioturbation can be described using a formulation which is similar to the one of diffusion / dispersion processes.

Transport of a component due to bioturbation can therefore be described as:

$$\underline{J}_{bt} = -\phi \rho_{mol} \underline{\underline{D}}_{bt} \, grad \, x^c \tag{3.21}$$

with

$$\underline{\underline{D}}_{bt} = \begin{pmatrix} D_{bt} & 0 \\ 0 & D_{bt} \end{pmatrix} \tag{3.22}$$

The size of D_{bt} at a certain depth in the sediment column depends on the number of organisms living in the sediment and on their activity. It is obvious, that D_{bt} decreases towards zero below a certain depth, as most of the organisms are living in the upper part of the sediment column. According to FOSSING et al. (2004), "it is not uncommon, that bioturbation in the upper centimeters of the seabed is as significant as molecular diffusion." The same coefficient D_{bt} can be applied to all components, as the activity of the organisms does not differ for different components (compare to FOSSING et al., 2004).

In *DiaTrans* the bioturbation coefficient D_{bt} is determined in the following way according to FOSSING et al. (2004):

$$D_{bt} = \begin{cases} 3.51 \cdot 10^{-10} \, m^2/s & b \leq 0.118 \, m \\ 3.51 \cdot 10^{-10} e^{-0.378(b-0.118)} \, m^2/s & b > 0.118 \, m \end{cases} \tag{3.23}$$

with b being the depth in the sediment column starting at the sediment-water interface. A function of the bioturbation coefficient (based on the formulation after FOSSING et al., 2004) vs. depth is qualitatively shown in fig. 3.8.

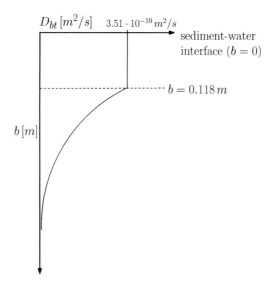

Figure 3.8: Function of bioturbation coefficient vs. depth (based on formulation after FOSSING et al., 2004)

The implementation and discretization of eq. 3.21 in the FVM is explained in detail in sec. 4.6.

3.8 Bioirrigation

In the uppermost fauna-inhabited sediment layers also the phenomenon of irrigating animals, which play an important role in the exchange of the solutes between sediment and bottom water, can be observed in nature. Those animals live in small tubes or small channels, through which solutes can easily and quickly be transported between the sediment and bottom water.

According to SCHLÜTER et al. (2000), bioirrigation results from the animal activity as they exchange bottom and burrow water for food supply or to flush their burrows of metabolic by-products (for more information see also ALLER, 1980; CHRIS-

3 Model Concepts, Physics & Biogeochemistry in Near-Shore Sediments

TENSEN et al., 1984; EMERSON et al., 1984; BOUDREAU and MARINELLI, 1994; BOUDREAU, 1997).

Compared to bioturbation (sec. 3.7) and diffusion (sec. 3.5), the transport via bioirrigation is much faster, as the solutes do not have to pass all sediment layers as they take a shortcut through the channels and tubes (compare to FOSSING et al., 2004).

In *DiaTrans*, transport due to bioirrigation is described as shown in eq. 3.24 according to SCHLÜTER et al. (2000), where x^c is the pore water concentration and x_0^c the bottom water concentration at the sediment-water interface:

$$B_i = \alpha_{bi1} e^{(-\alpha_{bi2} b)} \phi \rho_{mol} (x_0^c - x^c) \qquad (3.24)$$

As bioirrigation depends largely on the number of species living in the sediment and their activity, it is clear that it decreases towards zero with depth b (see fig. 3.9).

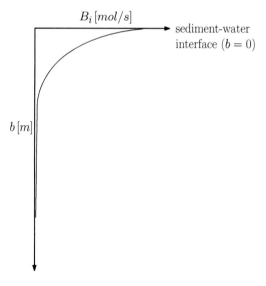

Figure 3.9: Function of bioirrigation term vs. depth (based on formulation after SCHLÜTER et al., 2000)

The irrigation parameters α_{bi1} and α_{bi2}, which vary seasonally, can be obtained using fitting techniques of field measurements. According to SCHLÜTER et al. (2000), $0 \leq \alpha_{bi1} \leq 10^{-5} \frac{1}{s}$ and $0 \leq \alpha_{bi2} \leq 50 \frac{1}{m}$. A function of the bioirrigation term (based on the formulation after SCHLÜTER et al., 2000) vs. depth is qualitatively shown in fig. 3.9.

For further information about bioirrigation see SCHLÜTER et al. (2000) or FOSSING et al. (2004), with the latter one offering another possibility to calculate the transport processes due to bioirrigation.

A description for the implementation and discretization of the bioirrigation term in the *FVM* is given in sec. 4.7.

3.9 Reactions

The consideration of diagenetic (biogeochemical) reaction processes is one of the main features in *DiaTrans*. Generally, as many reactions for as many components as necessary for certain applications can be accounted for in the model. For simplicity reasons, only the following dissolved components and some of their associated reactions are considered for the verification of the reaction processes (sec. 5.3) and their applications (sec. 6.2 and sec. 6.3). Note, that due to the multi-component formulation, the explicit consideration of solid components is not possible as only the liquid phase is modeled. The dissolved components included are:

- oxygen: O_2 with $M^{O_2} = 32.0\,g/mol$
- nitrate: NO_3 with $M^{NO_3} = 62.0\,g/mol$
- sulfate: SO_4 with $M^{SO_4} = 96.0\,g/mol$
- sulfide: H_2S with $M^{H_2S} = 32.4\,g/mol$
- manganese: Mn^{2+} with $M^{Mn^{2+}} = 54.9\,g/mol$
- iron: Fe^{2+} with $M^{Fe^{2+}} = 55.8\,g/mol$
- methane: CH_4 with $M^{CH_4} = 16.0\,g/mol$

Furthermore, pure water (H_2O with $M^{H_2O} = 18.0\,g/mol$) has to be included as the main component of the liquid phase and if density-dependent simulations have to be carried out, also chloride (Cl with $M^{Cl} = 35.4\,g/mol$) has to be incorporated as a component, although both of these components do not influence the governing primary and secondary redox reactions directly.

If other components than mentioned above should be accounted for in certain applications, the user is completely flexible with regard to the selection of reactions to fit the model to his needs. The incorporation of additional reactions can be done easily in

DiaTrans due to its object-oriented nature and the general formulation of the reaction terms. For further information which reaction processes have to be included for certain components (e. g. ammonia) see BOUDREAU (1996), WANG and VAN CAPPELLEN (1996) or JOURABCHI et al. (2005).

For certain reactions, concentrations of solid species such as organic matter (OM), solid manganese (MnO_2) or solid iron ($Fe(OH)_3$) are needed in order to calculate reaction rates. In this case, time independent profiles are defined which are either constant over the depth of the sediment column or they are decreasing exponentially over depth. A qualitative example of such profiles is given in fig. 3.10. In the verification and application chapters, the profiles are also quantitatively explained as they are used for the applications.

In the following, the governing reactions are explained in detail as well as they consideration as *governing component reaction balance equations*. Also detailed information about the reaction kinetics is given.

3.9.1 Governing Reactions

Generally, the governing reactions considered in *DiaTrans* can be split into primary and secondary redox reactions (**red**uction-**ox**idation reactions). Both of them can generally be described as "slow kinetically controlled reactions" (BOUDREAU, 1996). In redox reactions, atoms have their oxidation number (oxidation state) changed. Oxidation can be defined as an increase in oxidation number and reduction as a decrease in oxidation number. In case of the primary redox reactions, which are the basic reactions for OM oxidation, organic matter is the driving force for each reaction. The secondary redox reactions involve by-products and oxidants.

In all reactions listed below, also other components than the ones considered occur as the equations are written in a stoichiometric manner. Later, in the *governing component reaction balance equations* it is accounted for the choice of components which should actually be regarded in certain simulations.

In the stoichiometric eq. 3.25 to eq. 3.35, the explicitly simulated components are written in **bold** letters throughout and solid species with only assumed concentration profiles are written in ***bold symbol*** letters. $(CH_2O)_A(NH_3)_B(H_3PO_4)_C$ stands for total organic matter OM, MnO_2 for solid manganese and $Fe(OH)_3$ for solid iron. Note, that (H_2O) is not modeled explicitly, as the sum of all mole fractions always equals 1 in the multi-component formulation (see eq. 3.62).

Primary Redox Reactions

The degradation of organic matter (OM) is described by the so called "Primary Reactions" (eq. 3.25 to 3.30), which are reported by different authors (e.g. FROELICH et al., 1979; EMERSON et al., 1980; TROMP et al., 1995; BOUDREAU, 1996; EMERSON and HEDGES, 2003). FOSSING et al. (2004) states, that the order of the first five reactions reflects "(...) the energy obtained by the bacteria by degrading organic matter using oxygen, nitrate, [solid] manganese, [solid] iron and sulphate". Therefore, the degradation process by oxygen is the respiration process which yields most of the energy, whereas the process of sulfate respiration yields least of the energy.

This can also be seen, when looking at the concentration profiles in near-shore sediments, as those oxidants (oxygen, nitrate, manganese, iron and sulfate) "are consumed in exactly that order" (FOSSING et al., 2004). The existence of this successive pattern has been shown in many studies (e.g. FOSSING et al., 2004; BOUDREAU, 1996; WANG and VAN CAPPELLEN, 1996). This also means, that e.g. the process of denitrification only starts when almost all oxygen has been consumed or that manganese is only respired when almost no nitrate is present any more.

Additionally to those five primary reactions, a sixth one is incorporated in the model - which is methanogenesis (eq. 3.30). This reaction process is also included by different other authors (e.g. FROELICH et al., 1979; EMERSON et al., 1980; TROMP et al., 1995; BOUDREAU, 1996; WANG and VAN CAPPELLEN, 1996; EMERSON and HEDGES, 2003; JOURABCHI et al., 2005).

In those reaction equations, the coefficients A, B and C represent the *Redfield* numbers from a specified *Redfield ratio* $(A : B : C)$, which is $(106 : 16 : 1)$ for coastal sediments (e.g. BOUDREAU, 1996; WANG and VAN CAPPELLEN, 1996; EMERSON and HEDGES, 2003). The *Redfield ratio* is the molecular ratio of carbon, nitrogen and phosphorus in phytoplankton.

To obtain reaction rate terms for the primary redox reactions, *Monod*-type kinetics are adopted. The nature of these kinetics is described in detail in sec. 3.9.3.

Oxic Respiration

$$(CH_2O)_A(NH_3)_B(H_3PO_4)_C + (A+2B)(O_2)$$
$$\longrightarrow A(CO_2) + B(HNO_3) + C(H_3PO_4) + (A+B)(H_2O) \quad (3.25)$$

Denitrification

$$(CH_2O)_A(NH_3)_B(H_3PO_4)_C + \frac{4}{5}A(NO_3^-)$$
$$\longrightarrow \frac{1}{5}A(CO_2) + \frac{2}{5}A(N_2) + \frac{4}{5}A(HCO_3^-) + B(NH_3) + C(H_3PO_4) + \frac{3}{5}A(H_2O) \quad (3.26)$$

Mn(IV) Reduction

$$(CH_2O)_A(NH_3)_B(H_3PO_4)_C + 2A(MnO_2) + 3A(CO_2) + A(H_2O)$$
$$\longrightarrow 2A(Mn^{2+}) + 4A(HCO_3^-) + B(NH_3) + C(H_3PO_4) \quad (3.27)$$

Fe(III) Reduction

$$(CH_2O)_A(NH_3)_B(H_3PO_4)_C + 4A(Fe(OH)_3) + 7A(CO_2)$$
$$\longrightarrow 4A(Fe^{2+}) + 8A(HCO_3^-) + B(NH_3) + C(H_3PO_4) + 3A(H_2O) \quad (3.28)$$

Sulfate Reduction

$$(CH_2O)_A(NH_3)_B(H_3PO_4)_C + \frac{1}{2}A(SO_4^{2-})$$
$$\longrightarrow \frac{1}{2}A(H_2S) + A(HCO_3^-) + B(NH_3) + C(H_3PO_4) \quad (3.29)$$

Methanogenesis

$$(CH_2O)_A(NH_3)_B(H_3PO_4)_C$$
$$\longrightarrow \frac{1}{2}A(CH_4) + \frac{1}{2}A(CO_2) + (B-2C)(HCO_3^-) + B(NH_3) + C(H_3PO_4) \quad (3.30)$$

Secondary Redox Reactions

Eq. 3.31 to 3.35 describe the secondary redox reactions for the before mentioned components which are included for verification (sec. 5.3) and the applications (sec. 6.2 and 6.3). As mentioned before, without doubt, there are many more important secondary reactions which could be included in the model very easily, as *DiaTrans* is very flexible due to its object-oriented nature. BOUDREAU (1996) and WANG and VAN CAPPELLEN (1996) give more information about additional secondary redox reactions:

Fe^{2+} Re-Oxidation by O_2

$$4(\mathbf{Fe^{2+}}) + (\mathbf{O_2}) + 8(HCO_3^-) + 2(H_2O) \longrightarrow 4(\boldsymbol{Fe(OH)_3}) + 8(CO_2) \qquad (3.31)$$

Mn^{2+} Re-Oxidation by O_2

$$2(\mathbf{Mn^{2+}}) + (\mathbf{O_2}) + 4(HCO_3^-) \longrightarrow 2(\boldsymbol{MnO_2}) + 4(CO_2) + 2(H_2O) \qquad (3.32)$$

Sulfide Re-Oxidation by O_2

$$(\mathbf{H_2S}) + 2(\mathbf{O_2}) + 2(HCO_3^-) \longrightarrow (\mathbf{SO_4^{2-}}) + 2(CO_2) + 2(H_2O) \qquad (3.33)$$

Methane Re-Oxidation by O_2

$$(\mathbf{CH_4}) + 2(\mathbf{O_2}) \longrightarrow (CO_2) + 2(H_2O) \qquad (3.34)$$

Methane Re-Oxidation by SO_4^{2-}

$$(\mathbf{CH_4}) + (CO_2) + (\mathbf{SO_4^{2-}}) \longrightarrow 2(HCO_3^-) + (\mathbf{H_2S}) \qquad (3.35)$$

For the secondary redox reactions, "second-order"-type kinetics are used to obtain the reaction rate terms. The underlying principle is also described in sec. 3.9.3.

3.9.2 Governing Component Reaction Balance Equations

Governing component reaction balance equations can be obtained for each component which should be included in the model from eq. 3.25 to eq. 3.35. For each reaction which should be considered, a reaction rate term R_i can be defined, always using letters as indices which denote a certain reaction process. The sum of the reaction rate terms for each component $(\sum R_i)^c$ yields the reaction term r^c for the total balance equations (eq. 3.61 and 3.63), when it is multiplied with the porosity:

$$r^c = \phi (\sum R_i)^c \tag{3.36}$$

When multiplying $(\sum R_i)^c$ with the porosity, all terms in eq. 3.37 – 3.43 resulting from secondary reactions are multiplied with the effective porosity ϕ, as those secondary reactions are governed by the available pore space. For terms resulting from primary reactions, which are driven by solid organic matter concentrations, this multiplication leads to a multiplication with the solid porosity ϕ_s, as organic carbon is only available in the solid matrix of the sediment.

Below, the governing component reaction balance equations as they are considered in *DiaTrans* are listed, based on a formulation first presented by BOUDREAU (1996). Again, the coefficients A, B and C represent the *Redfield* numbers from a specified *Redfield ratio* $(A : B : C)$. As mentioned before, it is certainly possible to extend those balance equations to fit certain needs or to reduce them when less reactions should be included. When doing this, always the inter-dependency of different reactions must be checked, so that a realistic biogeochemical representation of the processes is guaranteed. A larger extent of balance equations is given in BOUDREAU (1996):

Oxygen (solute)

$$(\sum R_i)^{O_2} = -\frac{\phi_s(A+2B)}{\phi A}\{\sum k_i G_i\} R_{O_2} - \{2R_{SOx} + R_{MnOx} + R_{FeOx} + 2R_{CH_4Ox}\} \tag{3.37}$$

Nitrate (solute)

$$(\sum R_i)^{NO_3} = -\frac{\phi_s}{\phi}\{\sum k_i G_i\}\left(\frac{4}{5}R_{NO_3} - \frac{B}{A}R_{O_2}\right) \tag{3.38}$$

Sulfate (solute)

$$\left(\sum R_i\right)^{SO_4} = -\frac{\phi_s}{2\phi}\left\{\sum k_i G_i\right\} R_{SO_4} + R_{SOx} - R_{CH_4SO_4} \tag{3.39}$$

Sulfide (solute)

$$\left(\sum R_i\right)^{H_2S} = +\frac{\phi_s}{2\phi}\left\{\sum k_i G_i\right\} R_{SO_4} - R_{SOx} + R_{CH_4SO_4} \tag{3.40}$$

Manganese (solute)

$$\left(\sum R_i\right)^{Mn^{2+}} = +\frac{2\phi_s}{\phi}\left\{\sum k_i G_i\right\} R_{MnO_2} - 2R_{MnOx} \tag{3.41}$$

Iron (solute)

$$\left(\sum R_i\right)^{Fe^{2+}} = +\frac{4\phi_s}{\phi}\left\{\sum k_i G_i\right\} R_{Fe(OH)_3} - 4R_{FeOx} \tag{3.42}$$

Methane (solute)

$$\left(\sum R_i\right)^{CH_4} = +\frac{\phi_s}{2\phi}\left\{\sum k_i G_i\right\} R_{CH_4} - R_{CH_4Ox} - R_{CH_4SO_4} \tag{3.43}$$

In the equations above, G_i is the conventional symbol for organic matter concentration and k_i the reaction rate constants for organic matter decomposition. Usually it is assumed that there is a *reactive* fraction, k_1, of organic matter, that "decays within the top 10-20 cm of sediments, a more *refractory* component, k_2, that oxidizes on a scale of about ten to one hundred times longer, and a third that is largely *inert*" (BOUDREAU, 1997).

In the case of *DiaTrans*, the distribution of organic matter is described using a constant profile which is exponentially decreasing with depth, leading to the fact that the characteristics of those different fractions is already accounted for in the exponential distribution. Therefore, only a single rate constant for organic matter decomposition (k) is used in *DiaTrans*. An qualitative example of the organic matter concentration distribution can be seen in fig. 3.10.

The implementation and discretization of eq. 3.36 in the *FVM* is explained in detail in sec. 4.8.

3.9.3 Reaction Kinetics

The reaction rate terms "R_i" in equations 3.37 to 3.43 represent "the effect of oxidant concentrations and inhibition on the rate of OM decomposition" (BOUDREAU, 1996) or the kinetics of the different secondary reactions. In the following, detailed information is given about the adopted kinetic rate laws for the primary and secondary redox reactions.

Monod-Type Kinetics for Primary Reactions

For the primary redox reactions (eq. 3.25 to 3.30) *Monod*-type kinetics are usually adopted (BOUDREAU and WESTRICH, 1984; BOUDREAU, 1996, 1997). A general formulation for *Monod* kinetics (according to BOUDREAU, 1997) is presented in eq. 3.44:

$$\left(\frac{dG}{dt}\right)_{reaction} = -kG\frac{Ox}{K_{Ox} + Ox} \tag{3.44}$$

In this equation, K_{Ox} is a saturation constant and according to BOUDREAU (1997) this expression states that "the reaction is first order in the organic matter concentration $[G]$ and hyperbolic in the concentration of the oxidant $[Ox]$. The latter means that the reaction is essentially independent of Ox when $Ox \gg K_{Ox}$, and it is first order in Ox when $Ox \ll K_{Ox}$."

According to BOUDREAU (1997), inhibition "is the suppression of a reaction by the presence of another particular species." To represent inhibition mathematically, HUMPHREY (1972) and BLANCH (1981) offer different possible expressions employing an inhibition constant K'_I and I as the concentration of the inhibiting constituent. One of them is presented in eq. 3.45:

$$\left(\frac{dG}{dt}\right)_{reaction} = -kG\left(\frac{Ox}{K_{Ox} + Ox}\right)\left(\frac{K'_I}{K'_I + I}\right) \tag{3.45}$$

In the works of BOUDREAU (1996) and VAN CAPPELLEN et al. (1993) eq. 3.45 is successfully utilized to model organic matter decay in sediments and the rates of OM oxidation are formulated as follows (eq. 3.46 to eq. 3.51). Using this kind of formulation, the reactions further below in this sequence are "suppressed by the presence of unused oxidants involved in the previous reactions" (BOUDREAU, 1997).

3.9 Reactions

Below, the reaction rate terms for the primary reactions are listed, where the index 0 denotes the concentration of a constituent in the bottom water at the sediment-water interface. In eq. 3.46 to eq. 3.51 a term such as "(O_2)" denotes a molar concentration of a component (see eq. 3.6), so that $(O_2) = C_{mol}^{O_2} = \rho_{mol} \, x^{O_2}$. This definition applies for all components. The concentrations for the solid species MnO_2 and $Fe(OH)_3$, which are not explicitly modeled, are directly given as C_{mol} as their mole fractions x^c are not calculated during simulation. A qualitative distribution of those solid concentrations can be seen in fig. 3.10.

$$R_{O_2} = \frac{(O_2)}{K_{O_2} + (O_2)} = \frac{(x^{O_2}\rho_{mol})}{K_{O_2} + (x^{O_2}\rho_{mol})} \tag{3.46}$$

$$R_{NO_3} = \frac{(NO_3)}{K_{NO_3} + (NO_3)} \frac{K'_{O_2}}{K'_{O_2} + (O_2)} = \frac{(x^{NO_3}\rho_{mol})}{K_{NO_3} + (x^{NO_3}\rho_{mol})} \frac{K'_{O_2}}{K'_{O_2} + (x^{O_2}\rho_{mol})} \tag{3.47}$$

$$R_{MnO_2} = \frac{(MnO_2)}{(MnO_2)_0} \frac{K'_{NO_3}}{K'_{NO_3} + (NO_3)} \frac{K'_{O_2}}{K'_{O_2} + (O_2)} = \frac{C_{mol}^{MnO_2}}{C_{mol,0}^{MnO_2}} \frac{K'_{NO_3}}{K'_{NO_3} + (x^{NO_3}\rho_{mol})} \frac{K'_{O_2}}{K'_{O_2} + (x^{O_2}\rho_{mol})} \tag{3.48}$$

$$R_{Fe(OH)_3} = \frac{(Fe(OH)_3)}{(Fe(OH)_3)_0} \frac{K'_{O_2}}{K'_{O_2} + (O_2)} \frac{K'_{NO_3}}{K'_{NO_3} + (NO_3)} \frac{1}{1 + \frac{K'_{MnO_2}(MnO_2)}{(MnO_2)_0}} = \frac{C_{mol}^{Fe(OH)_3}}{C_{mol,0}^{Fe(OH)_3}} \frac{K'_{O_2}}{K'_{O_2} + (x^{O_2}\rho_{mol})} \frac{K'_{NO_3}}{K'_{NO_3} + (x^{NO_3}\rho_{mol})} \frac{1}{1 + \frac{K'_{MnO_2} C_{mol}^{MnO_2}}{C_{mol,0}^{MnO_2}}} \tag{3.49}$$

$$R_{SO_4} = \frac{(SO_4)}{K_{SO_4} + (SO_4)} \frac{K'_{O_2}}{K'_{O_2} + (O_2)} \frac{K'_{NO_3}}{K'_{NO_3} + (NO_3)} \frac{1}{1 + \frac{K'_{MnO_2}(MnO_2)}{(MnO_2)_0}} \frac{1}{1 + \frac{K'_{Fe(OH)_3}(Fe(OH)_3)}{(Fe(OH)_3)_0}} = \frac{(x^{SO_4}\rho_{mol})}{K_{SO_4} + (x^{SO_4}\rho_{mol})} \frac{K'_{O_2}}{K'_{O_2} + (x^{O_2}\rho_{mol})} \frac{K'_{NO_3}}{K'_{NO_3} + (x^{NO_3}\rho_{mol})} \frac{1}{1 + \frac{K'_{MnO_2} C_{mol}^{MnO_2}}{C_{mol,0}^{MnO_2}}} \frac{1}{1 + \frac{K'_{Fe(OH)_3} C_{mol}^{Fe(OH)_3}}{C_{mol,0}^{Fe(OH)_3}}} \tag{3.50}$$

$$R_{CH_4} = \frac{K'_{O_2}}{K'_{O_2} + (O_2)} \frac{K'_{NO_3}}{K'_{NO_3} + (NO_3)} \frac{1}{1 + \frac{K'_{MnO_2}(MnO_2)}{(MnO_2)_0}}$$

$$\frac{1}{1 + \frac{K'_{Fe(OH)_3}(Fe(OH)_3)}{(Fe(OH)_3)_0}} \frac{K'_{SO_4}}{K'_{SO_4} + (SO_4)} =$$

$$\frac{K'_{O_2}}{K'_{O_2} + (x^{O_2}\rho_{mol})} \frac{K'_{NO_3}}{K'_{NO_3} + (x^{NO_3}\rho_{mol})} \frac{1}{1 + \frac{K'_{MnO_2} C^{MnO_2}_{mol}}{C^{MnO_2}_{mol,0}}}$$

$$\frac{1}{1 + \frac{K'_{Fe(OH)_3} C^{Fe(OH)_3}_{mol}}{C^{Fe(OH)_3}_{mol,0}}} \frac{K'_{SO_4}}{K'_{SO_4} + (x^{SO_4}\rho_{mol})} \quad (3.51)$$

Kinetics for Secondary Reactions

VAN CAPPELLEN and WANG (1995, 1996) have done an admirable job of collecting information about the secondary reactions and also suggested kinetic rate laws for them. These kinetics for the secondary reactions (eq. 3.31 to 3.35) were adopted and are presented below, as they are used in *DiaTrans* for the considered components mentioned above. These reactions can be described as "second order chemical reaction" with two species reacting irreversibly at a rate constant k_j. As for the primary reactions, a term such as "(O_2)" denotes a molar concentration of a component.

$$R_{MnOx} = k_{MnOx}(Mn^{2+})(O_2) = k_{MnOx}(x^{Mn^{2+}}\rho_{mol})(x^{O_2}\rho_{mol}) \quad (3.52)$$

$$R_{FeOx} = k_{FeOx}(Fe^{2+})(O_2) = k_{FeOx}(x^{Fe^{2+}}\rho_{mol})(x^{O_2}\rho_{mol}) \quad (3.53)$$

$$R_{SOx} = k_{SOx}(H_2S)(O_2) = k_{SOx}(x^{H_2S}\rho_{mol})(x^{O_2}\rho_{mol}) \quad (3.54)$$

$$R_{CH_4Ox} = k_{CH_4Ox}(CH_4)(O_2) = k_{CH_4Ox}(x^{CH_4}\rho_{mol})(x^{O_2}\rho_{mol}) \quad (3.55)$$

$$R_{CH_4SO_4} = k_{CH_4Ox}(CH_4)(SO_4) = k_{CH_4Ox}(x^{CH_4}\rho_{mol})(x^{SO_4}\rho_{mol}) \quad (3.56)$$

3.9 Reactions

Parameters and Rate Constants

Table 3.3 lists parameters and kinetic rate constants for the secondary reactions, as well as *Monod* and inhibition constants, as they are used in *DiaTrans*. It is easily possible to use different parameters due to the flexible nature of the model.

Table 3.3: Biogeochemical parameters and rate constants for coastal zones

Condition or Rate Constant	Value	Units	Value used in DiaTrans	Units used in DiaTrans	Sources
$RedfieldRatio$ $(A:B:C)$	106:16:1	-	106:16:1	-	BOUDREAU (1996)
k	1	$1/year$	3×10^{-8}	$1/s$	BOUDREAU (1996)
K_{O_2}	0.008	$mmol/L$	0.008	mol/m^3	VAN CAPPELLEN and WANG (1996)
K'_{O_2}	0.008	$mmol/L$	0.008	mol/m^3	VAN CAPPELLEN and WANG (1996)
K_{NO_3}	0.030	$mmol/L$	0.030	mol/m^3	JAHNKE et al. (1982)
K'_{NO_3}	0.030	$mmol/L$	0.030	mol/m^3	JAHNKE et al. (1982)
K_{SO_4}	1.000	$mmol/L$	1.000	mol/m^3	BOUDREAU and WESTRICH (1984)
K'_{SO_4}	1.000	$mmol/L$	1.000	mol/m^3	BOUDREAU and WESTRICH (1984)
K'_{MnO_2}	10	-	10	-	BOUDREAU (1996)
$K'_{Fe(OH)_3}$	10	-	10	-	BOUDREAU (1996)
k_{MnOx}	2×10^6	$L/(mmol\ y)$	0.06342	$m^3/(mol\ s)$	VAN CAPPELLEN and WANG (1995)
k_{FeOx}	2×10^6	$L/(mmol\ y)$	0.06342	$m^3/(mol\ s)$	VAN CAPPELLEN and WANG (1995)
k_{SOx}	6×10^5	$L/(mmol\ y)$	0.01903	$m^3/(mol\ s)$	VAN CAPPELLEN and WANG (1995)
k_{CH_4Ox}	1×10^7	$L/(mmol\ y)$	0.3171	$m^3/(mol\ s)$	VAN CAPPELLEN and WANG (1995)
$k_{CH_4SO_4}$	1×10^7	$L/(mmol\ y)$	0.3171	$m^3/(mol\ s)$	VAN CAPPELLEN and WANG (1995)

As mentioned before, constant distribution profiles for the solid species which have to be considered (i.e. OM, MnO_2 and $Fe(OH)_3$) must be given in *DiaTrans*. For OM, MnO_2 and $Fe(OH)_3$ profiles are generally chosen, which exponentially decrease with depth. The shape of these profiles has been reported by e.g. BOUDREAU (1996);

WANG and VAN CAPPELLEN (1996); JOURABCHI et al. (2005). A qualitative description of those profiles is shown in fig. 3.10. Note, that the solid concentrations are expressed per unit volume total sediment.

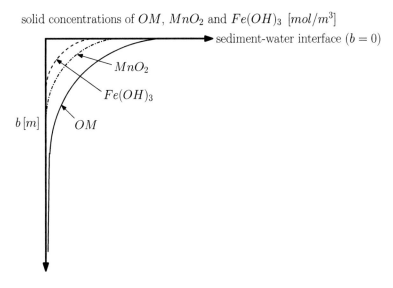

Figure 3.10: Qualitative distribution of solid concentrations (OM, MnO_2 and $Fe(OH)_3$)

3.10 The Multi-Component Formulation

Generally, the governing equations that describe one-phase / multi-component flow, transport and biogeochemical processes in the subsurface can be written in different differential formulations so that the coupling of those equations can be handled appropriately to the task to be performed.

DiaTrans uses a *fully or direct coupled formulation*, in which the governing equations for density-dependent flow, transport and reactions including bioturbation and bioirrigation are treated as fully coupled simultaneous equations. Each component (e.g. pure water, chloride or methane) yields one governing equation, with all of them being summarized in one system of equations. After discretization, for every time-step a highly nonlinear sparse system of algebraic equations is solved in a fully-implicit procedure for the primary variables *pressures of the water phase* and *mole fractions* of the considered components (see also SCHANKAT et al., 2008a,b, 2009a).

This fully-implicit approach for all of the above mentioned processes is unique in the field of subsurface flow and transport modeling, as most of the present models only solve for either flow and transport or reactions and often neglect density dependency. Decoupled formulations such as *operator-splitting* or *iterative two-step* methods are also often used, although they have major disadvantages compared to the direct coupled formulation as outlined in sec. 1.1.2. One of these major disadvantages is that different time steps have to be used for flow, transport and reactions which then again have to be adjusted to each other in the course of the computation.

In the following sections it is explained how the governing equations for each component are obtained, namely the continuity equation, the momentum equation and the equation for the density dependency. All of the physical and biogeochemical processes taken into account in the continuity equation are described in sec. 3.4 to 3.9.

3.10.1 Continuity Equation

In the following, a summarized overview is presented, which describes how the continuity equation for a single phase multi-component system in porous media is obtained. The descriptions presented herein are mostly obtained from HINKELMANN (2005) who also gives more detailed explanations.

3 Model Concepts, Physics & Biogeochemistry in Near-Shore Sediments

The so-called general form of the balance equations can be used to formulate the continuity equation. A fixed *Eulerian* control volume Ω is considered with an extensive state variable $E(t)$, "which is the volume integral of a scalar or vector entity $e\,[\underline{x}(x,y,z),t]$. $\underline{x}$ denotes a spatial vector with the components x, y, z and t stands for time" (HINKELMANN, 2005).

$$E(t) = \int_{\Omega} e(\underline{x}, t) dV \qquad (3.57)$$

The general form of the balance equation states "that the volume integral of the temporal change of the entity e plus the fluxes $\underline{F}$ multiplied by the normal vector $\underline{n}$ and integrated over the surface $\Gamma = \partial \Omega$ minus the volume integral of sink or source terms $r(\underline{x}, t)$ equals zero (see fig. 3.11). According to HINKELMANN (2005), "the fluxes consist of an advective part $\underline{v}(\underline{x},t)e(\underline{x},t)$ and a diffusive part $\underline{w}(\underline{x},t)$" (see eq. 3.58).

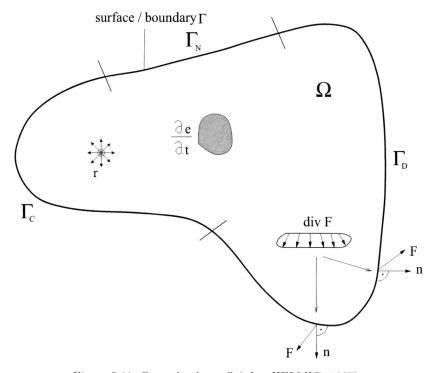

Figure 3.11: Control volume Ω (after HELMIG, 1997)

3.10 The Multi-Component Formulation

HINKELMANN (2005) further describes, how the general balance equation reads in integral form (eq. 3.58) and how the surface integral can be transferred into a volume integral applying the *Green-Gauss Integral Theorem* (see eq. 3.59):

$$\int_\Omega \frac{\partial e}{\partial t} dV + \int_\Gamma (\underline{v}e + \underline{w})\underline{n} dO - \int_\Omega r dV = 0 \qquad (3.58)$$

$$\int_\Omega \left[\frac{\partial e}{\partial t} + div\,(\underline{v}e + \underline{w}) - r \right] dV = 0 \qquad (3.59)$$

"As this equation can be applied to arbitrary control volumes, the integral form can be replaced by a differential form assuming a continuous integrand" (HINKELMANN, 2005). The resulting equation is presented below.

$$\frac{\partial e}{\partial t} + div\,(\underline{v}e + \underline{w}) - r = 0 \qquad (3.60)$$

In order to solve an initial boundary-value problem, initial conditions for all primary variables must be specified in the entire domain Ω, as well as boundary conditions along the boundary Γ (HINKELMANN, 2005). For more detailed information about initial and boundary conditions see sec. 4.10, HELMIG (1997) or HINKELMANN (2005).

Generally, multi-component formulations are often utilized in multi-phase flow and transport models, e.g. *MUFTE_UG* (see HELMIG, 1997; HELMIG et al., 1998; HINKELMANN, 2005). In the one-phase / multi-component formulation as used in *DiaTrans* the detailed description of the continuity equation for the conservation of mass of each component c (eq. 3.61) is obtained by assuming that the porosity does not change with respect to time and by substituting the following terms into the general form of the balance equation (eq. 3.60):
$e = \phi \rho_{mol} x^c$, $\underline{v} = \underline{v}_a$, $\underline{w} = -div(\phi \rho_{mol} \underline{\underline{D}}_{hyd} grad\,x^c) - div(\phi \rho_{mol} \underline{\underline{D}}_{bt}\,grad\,x^c)$ and $r = \alpha_{bi1} e^{(-\alpha_{bi2} b)} \phi \rho_{mol} (x_0^c - x^c) + r^c + q^c$.

The continuity equation in conservative form then reads as:

$$\phi \frac{\partial(\rho_{mol} x^c)}{\partial t} + div(x^c \rho_{mol} \underline{v}) - div(\phi \rho_{mol} \underline{\underline{D}}_{hyd} grad\,x^c)$$
$$- div(\phi \rho_{mol} \underline{\underline{D}}_{bt}\,grad\,x^c) - \alpha_{bi1} e^{(-\alpha_{bi2} b)} \phi \rho_{mol} (x_0^c - x^c) - r^c - q^c = 0 \quad (3.61)$$

Again, from mass conservation it follows, that:

$$\sum_c x^c = 1 \qquad (3.62)$$

The terms presented in eq. 3.61, which represent the physical and biogeochemical processes in *DiaTrans*, are the following:

- advective flux (see sec. 3.4 & 4.4) $\underline{J}_a = div(x^c \rho_{mol} \underline{v})$
- diffusive / dispersive flux (see sec. 3.5 & 4.5) $\underline{J}_d = -div(\phi \rho_{mol} \underline{\underline{D}}_{hyd} grad\, x^c)$
- bioturbation flux (see sec. 3.7 & 4.6) $\underline{J}_{bt} = -div(\phi \rho_{mol} \underline{\underline{D}}_{bt}\, grad\, x^c)$
- bioirrigation term (see sec. 3.8 & 4.7) $B_i = \alpha_{bi1} e^{(-\alpha_{bi2} b)} \phi \rho_{mol}(x_0^c - x^c)$
- reaction term r^c (see sec. 3.9 & 4.8)

The first term in eq. 3.61 corresponds to the rate of accumulation $\phi \frac{\partial(\rho_{mol} x^c)}{\partial t}$ (see sec. 4.3) and the last term is the the source / sink term q^c (see sec. 4.9). With the above expressions, eq. 3.61 reads as:

$$\phi \frac{\partial(\rho_{mol} x^c)}{\partial t} + div\, \underline{J}_a + div\, \underline{J}_d + div\, \underline{J}_{bt} - B_i - r^c - q^c = 0 \qquad (3.63)$$

The numerical implementation and discretization of all terms in eq. 3.63 using the *Finite-Volume-Method* is given in sec. 4.2 to 4.9.

Note, that the model *MUFTE_UG* (e.g. HELMIG, 1997; HELMIG et al., 1998; HINKELMANN, 2005) utilizes mass fractions X^c instead of mole fractions x^c for each component. The detailed continuity equation in *MUFTE_UG* is explained in sec. 6.1.1.

In both cases the generalized *Darcy* law (eq. 3.64) serves as the momentum equation to obtain the *Darcy* velocity $\underline{v}$ (see sec. 3.10.2).

3.10.2 Momentum Equation

The momentum equation is replaced with the generalized *Darcy* law in subsurface systems (HINKELMANN, 2005) which yields the *Darcy* velocity $\underline{v}$ and reads as the following:

$$\underline{v} = -\frac{\underline{\underline{k}}}{\mu}(grad\, p - \rho_{mass}\underline{g}) \qquad (3.64)$$

This law was obtained carrying out experiments and can also be derived from the *Navier-Stokes* equations applying some assumptions and simplifications, such as neglecting the inertia terms. It is only valid for *Reynold*'s numbers less than 1. (HINKELMANN, 2005)

The *Darcy* velocity is an average velocity as it is assumed that the whole cross-sectional area of a domain can act as potential flow paths.

The actual pore velocity of fluid particles, also called *seepage* velocity $\underline{v}_a$, is obtained by dividing the *Darcy* velocity by the effective porosity as only a part of the control volume will contribute to the flow (i.e. only the volume which is filled with mobile fluid) (see eq. 3.65).

$$\underline{v}_a = \frac{\underline{v}}{\phi} \tag{3.65}$$

3.10.3 Density Dependency

As mentioned before, one of the main advantages of *DiaTrans*, compared to other flow and transport models for processes in near-shore sediments, is that it is possible to account for density dependent flow. The numerical modeling approach takes into account that freshwater and chloride enriched saltwater are miscible – in contrast to a sharp-interface approach (e.g. HUYAKORN et al., 1996), in which freshwater and saltwater are considered to be immiscible. Thus, in the approach with miscible fluids a mixing zone occurs where the salinity gradually changes due to hydrodynamic dispersion, and a one-phase/multi-component model concept is required (see OLDENBURG and PRUESS, 1995; KOLDITZ et al., 1998; HINKELMANN et al., 2000a). The consideration of the density dependency is vital as diffusion and dispersion processes are assumed to be dominant and the width of the transition zone cannot be neglected. Generally, the density of a fluid phase depends on different factors such as temperature, pressure or salinity. In *DiaTrans*, the fluid is regarded as being incompressible and only the dependence on the salinity (i.e. chloride concentration) is regarded explicitly.

To calculate the mass density ρ_{mass} of the water phase, which depends on the salinity S, a modified formulation after LANG (1990) is used. This has been suggested by HINKELMANN et al. (2002) and has been successfully applied in different fields of application.

After LANG (1990), the mass density of water is determined using a reference mass density of pure water (ρ_0) without dissolved chloride or other constituents in it and is calculated as follows with $T_0 = 4°C$ and ρ_0 equal to $999.972\,kg/m^3$:

$$\rho_{mass} = \rho_0 \left[1 - \left(7\,(T - T_0)^2 - 750\,S\right) \cdot 10^{-6}\right] \qquad (3.66)$$

As in HINKELMANN et al. (2002), it is assumed that "the volume of a given amount of water remains constant" when chloride or other constituents are solved in it. In *DiaTrans* the mass density is solely a function of chloride concentration as other constituents such as sulfate or methane have a negligible effect on the density as their concentrations are usually small compared to the ones of chloride.

MÜLLER (1999) presents a formula to calculate the salinity S from chlorinity Cl for seawater in the North and Baltic Sea:

$$(S‰) = 1.80665\,(Cl‰) \qquad (3.67)$$

To calculate the chlorinity from the mole fraction of chloride x^{Cl}, we use a modification of the relation according to HINKELMANN et al. (2002) with ρ_w being the constant water density of $1000\,g/L$ and the molecular weights being $M^{Cl} = 0.0354\,kg/mol$ for chloride and $M^{H_2O} = 0.018\,kg/mol$ for pure freshwater:

$$1\,(Cl‰) \approx \frac{((1g/L)/M^{Cl})}{(\rho_w/M^{H_2O} + (1g/L)/M^{Cl})} = 0.0005082 \qquad (3.68)$$

Eq. 3.68 states that a chlorinity value of $1‰$ corresponds to a chloride mole fraction of $x^{Cl} = 0.0005082$. Using this information and the assumption that the chloride concentrations are much smaller compared to the constant water density it follows that:

$$Cl = \frac{x^{Cl}}{0.0005082} \qquad (3.69)$$

Inserting eq. 3.67 and eq. 3.69 into eq. 3.66 yields the governing equation for the density dependency of the water phase depending on the chloride mole fraction x^{Cl} as it is considered in *DiaTrans* (eq. 3.70).

Usually, a water temperature of $T = 10°C$ is assumed in *DiaTrans*. Then, a chloride concentration of 0.0 mmol/L leads to a mass density equal to $999.72\,kg/m^3$.

$$\rho_{mass} = \rho_0 \left[1 - \left(7\,(T - T_0)^2 - 750 \cdot \frac{1.80665}{0.0005082} x^{Cl}\right) \cdot 10^{-6}\right] \qquad (3.70)$$

The density dependency of the water phase is represented differently in the formulation of *MUFTE_UG* (e.g. HELMIG, 1997; HELMIG et al., 1998; HINKELMANN, 2005), the model which is used in sec. 6.1, as *MUFTE_UG* uses mass fractions X^c for the two components saltwater and freshwater. A detailed description of this approach after OLDENBURG and PRUESS (1995) is presented in sec. 6.1.1.

4

Numerical Methods & Implementation

This chapter starts with a review about the advantages and the disadvantages of the programming language *Java*, which was chosen to develop the model *DiaTrans*. Then, the main aspects of the object-oriented implementation in *DiaTrans* are outlined, including an overview of the internal class structure.

Information about the numerical methods utilized, the implementation and discretization of all physical and biogeochemical terms in the multi-component formulation is given, including details about the semi-discrete *Finite-Volume-Method*, the upwinding technique and initial and boundary conditions.

Also, it is outlined how nonlinear systems of equations are solved using the *Newton-Raphson-Method*, with remarks on the inner linear solver and preconditioners. The chapter closes with information about the adaptive time-stepping technique as well as methods for pre- and postprocessing utilized throughout this work.

4.1 Object-Oriented Programming in Java

DiaTrans was developed using the object-oriented programming language *Java* in contrast to other models which use languages such as *C*. The choice of *Java* has different reasons. CHOUDHARI (2001) presents a good overview about the advantages and the main disadvantages of *Java* of which some of them are presented below. Only the benefits and disprofits concerning the field of application of *DiaTrans* are listed here, skipping items such as networking or multimedia.

4.1.1 Advantages

All of the major advantages have been summarized by the developers of *Java*, Sun Microsystems, who state that "Java is simple, object-oriented, distributed, interpreted, robust, secure, architecture neutral, portable, multi threaded and dynamic."

One of those benefits is, that *Java* is much "simpler" due to its object-oriented nature compared to popular programming languages such as *C*, *C++* or *FORTRAN*, as *Java* has "replaced the complexity of multiple inheritance in *C++* with a simple structure called interface and also has eliminated the use of pointers" (CHOUDHARI, 2001). It is also simple, as automatic memory allocation and garbage collection is used and because it provides a clean syntax which makes it easy to write and read.

Object-oriented programming simulates the real world, as everything can be modeled as objects which interact with each other and can easily be manipulated. *Java* also uses a tree-type hierarchy using object classes with child classes having the ability to inherit properties and behaviors of the parent class. This functionality and the structured programming technique in *Java* introduces a high amount of generalization in *DiaTrans*, which makes it easy to understand and extend the model. One single parent class can be used for each element of the *Finite-Volume* grid or one for the different physical or biogeochemical processes for example. Therefore, the general possibility to incorporate additional physical and biogeochemical processes, such as the *Darcy-Brinkmann* equation (e.g. KHALILI et al., 1999; SCHEUERMANN et al., 2001) or supplemental reaction kinetics, is given. The properties and the behaviour of the different dissolved components in the multi-component formulation are similar, which makes it also possible to easily use the model for as many components as necessary for a certain field of application without having to manipulate the code.

According to CHOUDHARI (2001) one of the most compelling reasons to model in *Java* is its "platform independence", as it runs on most major hardware and software platforms, including Windows, Macintosh or UNIX. As *DiaTrans* is designed to be used for further research projects at the *Alfred Wegener Institute for Polar and Marine Research*, Bremerhaven, Germany and at the *Chair of Water Resources Management and Modeling of Hydrosystems*, Technische Universität Berlin, Germany, the portability and platform independence is of great importance.

A further convenience of *Java* is its reliability. Reliability is accomplished by different means such as the avoidance of pointers to eliminate the possibility of overwriting memory or corrupting data, the automatic type conversion and the additional possibilities to check the code to identify type mismatches or other inconsistencies (see CHOUDHARI, 2001). It also removes certain types of programming constructs which are prone to errors.

4.1.2 Disadvantages

When talking about the use of *Java* to perform tasks in numerics, it is often stated that the main disadvantage is speed. This was especially true at the beginning of *Java* development until the end of the last century, as the methods to create the code were very inefficient. At those times, it was revealed that it was about one order of magnitude slower than the programming language *C*. This speed difference is mainly due to the fact, that "an interpreter must first translate the Java binary code into the equivalent microprocessor instructions" (CHOUDHARI, 2001), which obviously takes some time.

But already in the year 2003, LEWIS and NEUMANN (2003) presented a summary of benchmarks which show "that Java performance on numerical code is comparable to that of C++, with hints that Java's relative performance is continuing to improve." One of the benchmark tests occurring in the summary of LEWIS and NEUMANN (2003), is the one of BULL et al. (2003), which shows that "the performance gap between Java and more traditional scientific programming languages is no longer a wide gulf" and that "the performance gap is small enough to be of little or no concern to programmers" on Intel Pentium or UNIX machines (BULL et al., 2003). Additionally, LEWIS and NEUMANN (2003) presented benchmarks in which Java can already be a little bit faster than *C* and they suggest that *Java* performance is catching up to or even pulling ahead of *C*.

All this corresponds to the findings of sec. 6.1, showing that the modeling of physical processes at *sand boils*, with small and medium sized systems of equations, could be carried out approximately at the same computational speed when using *DiaTrans* (developed in *Java*) compared to *MUFTE_UG* (developed in *C*). Further improvements have been achieved when applying the adaptive time-stepping technique, which is presented in sec. 4.12.

4.1.3 Implementation in DiaTrans

In the following, the main characteristics of *DiaTrans* concerning the implementation of object-oriented programming features are outlined. The object-oriented programming approach provides useful properties for the implementation of the multi-component formulation on two-dimensional grids.

To set up the computational mesh for example, an object-oriented approach has certain advantages compared to other programming languages such as *C*. Nodes, edges or finite-volume elements in two-dimensional space can be considered as objects. A node for example is defined by its coordinates in the two dimensions, and can also have additional information associated with it, e.g. its ID. Although, two node objects can have different coordinates and IDs, they are similar to each other as they store the same kind of information and have the same kind of behaviour. Thus, they belong to the same class, which serves as a template to create any number of node instances that may be required by the program (see SINGH NOTAY, 2007). Therefore it is easy to account for variable domain sizes and mesh resolutions.

The same concept is applied for the components considered in *DiaTrans*. As each component is similar, it can be treated in the same way, e.g. the same kind of information is stored such as the molecular diffusion coefficient, the molecular weight or the characteristics for possible reactions. This approach is very useful in the multi-component formulation, as it can be used for an arbitrary number of components without manipulating the code.

As the nature of physical processes can be very similar as well (e.g. diffusion and bioturbation), these processes can also be handled as classes. This generalized programming approach makes *DiaTrans* very suitable for further development and the incorporation of additional processes or reaction kinetics, which are also handled as objects with similar attributes.

Packages can be used in object-oriented programming to group classes, which are similar or highly interact with each other. Below, the main packages to set up the computational mesh and to arrange the balance equations are listed (see SINGH NOTAY, 2007):

- ***geometry***: The geometry package contains the classes related to two-dimensional geometry, e.g. Node, Edge, etc.

4.1 Object-Oriented Programming in Java

- ***fvGeometry***: the fvGeometry package contains classes related with the Finite Volume Geometry, e.g. FvNode, FvEdge, etc. All these classes are extended from the corresponding classes in the geometry package for example.

- ***grid***: The *Finite-Volume* grid and the classes to generate grids are grouped in this package.

- ***model***: This package is made up of the classes and methods for the physical and biogeochemical processes to build the balance equations for the multi-component formulation. This includes advection, diffusion / dispersion, bioturbation, bioirrigation and the different reaction equations for primary and secondary reactions. Also the method to consider the density dependency belongs to this package.

  The classes to solve the system of discretized nonlinear multi-component equations are also incorporated in this package. The class *AbstractNewtonRaphsonEquationSolver*, for example, provides the template for the nonlinear *Newton-Raphson* method with the inner linear *BiCGSTAB* solver.

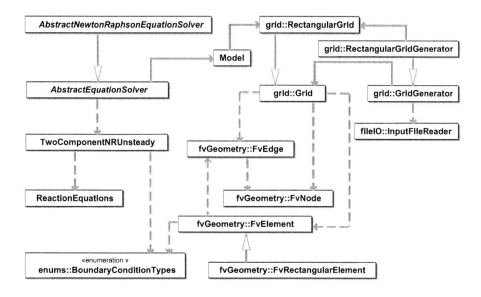

Figure 4.1: Simplified flow chart of *DiaTrans*

Further, the following two packages are employed for pre- and postprocessing.

- ***enums***: This package contains the enumeration types for the boundary conditions (DIRICHLET, NEUMANN), the scalar field types for the finite-volume cells (e. g. pressure, permeability, mole fraction) or the vector field types (*Darcy* and seepage velocity).

- ***fileIO***: This package contains the standard communication classes for pre- and postprocessing, i. e. the class for reading the input file from the *Input-File-Editor* as well as the class to write the output files for the visualization in *Paraview* (see also sec. 4.13.2).

A simplified flow chart including the main classes of *DiaTrans* is presented in fig. 4.1. For more information about the internal class structure see SINGH NOTAY (2007).

4.2 Finite-Volume-Method

For the discretization in the spatial coordinates *DiaTrans* utilizes the *Finite-Volume-Method (FVM)* which "has been established in engineering sciences for several decades" (HINKELMANN, 2005). Especially in water-related subjects and questions dealing with flow and transport processes, the *FVM* is of great importance. Almost all new model developments in this field utilize this technique. When the *FVM* is used for rectangular structured grids as in *DiaTrans*, this method is also called *Integral-Finite-Difference Method (IFDM)*. It may be used to solve problems for any arbitrary or structured grid. In this respect, it is similar to the *Finite-Element-Method* and makes it capable to solve physical problems for domains having a complex geometry.

A control volume is defined for every node of the computational mesh and may be chosen in more than one way. *DiaTrans* uses a *cell-centered FVM*, where the control volume is the element or cell and the physical quantities are computed at the center of the grid elements (see fig. 4.2, left). Another approach may be to calculate the unknown values at the nodes of the grid. This approach is called *node-centered*, having different possibilities of finding the control volume (fig. 4.2, middle and right) (see HINKELMANN, 2005).

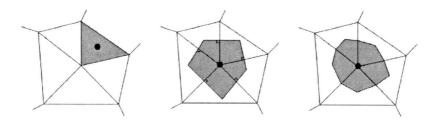

Figure 4.2: Control volumes of the FVM (modified after HINKELMANN, 2005)

When using the *FVM*, the differential forms of the balance equations in their conservative form are integrated over all control volumes. According to HINKELMANN (2005), in this way "a local conservation of the considered equation is guaranteed". For a general case, we may write the conservation equation (eq. 3.59) in the form:

$$\int_\Omega \frac{\partial e}{\partial t} dV + \int_\Omega div \underline{F} dV - \int_\Omega r dV = 0 \qquad (4.1)$$

4 Numerical Methods & Implementation

Again, with e being a scalar or vector entity, $\underline{F}$ being a flux term consisting of an advective and a diffusive / dispersive part, r being a sink/source term, and Ω being the solution domain of the control volume. As described in sec. 3.10.1, the volume integral in eq. 4.1 can be transferred into a surface integral applying the *Green-Gauss* Integral Theorem (see HINKELMANN, 2005) leading to eq. 4.2 with $\underline{n}$ being the normal vector.

$$\int_\Omega \frac{\partial e}{\partial t} dV + \int_\Gamma \underline{F}\,\underline{n}\,dO - \int_\Omega r\,dV = 0 \qquad (4.2)$$

In the case of *DiaTrans* we can rewrite the balance equations in their differential forms (eq. 3.61 and 3.63) as shown in eq. 4.3 and eq. 4.4 to obtain formulas similar as described in eq. 4.2:

$$\int_\Omega [\phi \frac{\partial(\rho_{mol} x^c)}{\partial t}] dV + \int_\Gamma [(x^c \rho_{mol}\underline{v}) - (\phi \underline{\underline{D}}_{hyd}\rho_{mol} grad\,x^c) - (\phi\rho_{mol}\underline{\underline{D}}_{bt}\,grad\,x^c)]\underline{n}\,dO$$
$$- \int_\Omega [\alpha_{bi1} e^{(-\alpha_{bi2} b)} \phi\rho_{mol}(x_0^c - x^c) + r^c + q^c] dV = 0 \qquad (4.3)$$

$$\int_\Omega [\phi \frac{\partial(\rho_{mol} x^c)}{\partial t}] dV + \int_\Gamma [\underline{J}_a + \underline{J}_d + \underline{J}_{bt}]\underline{n}\,dO - \int_\Omega [B_i + r^c + q^c] dV = 0 \qquad (4.4)$$

Each term in eq. 4.3 or eq. 4.4 (rate of accumulation, advective flux, diffusive / dispersive flux, bioturbation flux, bioirrigation term, reaction term, source / sink term) must be determined, using numerical techniques which are outlined for each term in sec. 4.3 to 4.9, including a detailed explanation of their discretized implementation in *DiaTrans*. Taking initial and boundary conditions for all primary variables into account, a solution is obtained.

Throughout the following explanations of the discretization (sec. 4.3 to 4.9), a standard compass notation is used (see fig. 4.3).

One inner control volume cell (**C**enter) is bounded by four interfaces (**n**orth, **e**ast, **s**outh and **w**est) and has four direct neighbor cells (**N**orth, **E**ast, **S**outh and **W**est). The width of one cell in simulated horizontal x-direction (e.g. distance between **W** and **C**) is throughout denoted as Δx, the width of a cell in simulated y-direction (e.g. distance between **C** and **N**) as Δy. As *DiaTrans* only uses structured rectangular meshes, Δx and Δy are always constant. In the special case of structured quadratic

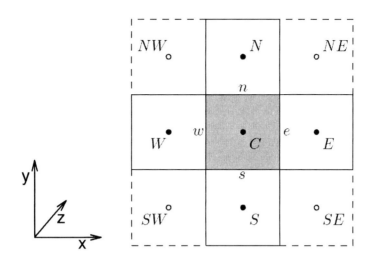

Figure 4.3: Compass notation for rectangular structured two-dimensional grid

meshes, Δx and Δy are equal. In a two-dimensional *FVM*, an interface (edge area) is defined as either $(\Delta x \cdot \Delta z)$ for the north and south interface or $(\Delta y \cdot \Delta z)$ for the east and west interface. The volume of a cell is defined as $(\Delta x \cdot \Delta y \cdot \Delta z)$, where Δz is a constant depth value of the cells in the third coordinate direction, with this direction not being computed in 2D. Δz is usually set to 1 m. Note that *DiaTrans* is usually simulating processes in a 2D vertical plane but can easily be adjusted to simulate 2D horizontal planes by neglecting all terms caused by gravity. Further, a boundary cell only has three neighbors and interfaces and a corner cell has only two neighbors and interfaces.

The surface integral in eq. 4.3 and eq. 4.4 describes the sum of fluxes across all four interfaces, whereas a volume integral in the above equations denotes a term which is integrated over the volume of a cell. Therefore, each control volume is *locally mass conservative* and consequently the *FVM* is also *globally mass conservative* over the whole computational domain.

4.3 Discretization of Rate of Accumulation

The accumulation term in eq. 3.63 is discretized using a time discretization scheme. Generally, the time domain is subdivided into a number of constant or variable time-

4 Numerical Methods & Implementation

steps Δt^n where the solution functions are determined. For the time t^0, the initial conditions for all primary variables must be given:

$$t^n = t^0 + n\Delta t^n \tag{4.5}$$

In this equation, t^n denotes the time after the n-th time step and n the number of time steps (HINKELMANN, 2005).

The time-dependent equations determined for the governing equations in sec. 3.10 and sec. 4.2 can be written in generalized form in the following way:

$$\frac{\partial e}{\partial t} = Ae \tag{4.6}$$

Here again, e stands for a scalar or vector entity, and the operator A (system matrix) contains the spatial derivatives as well as the other terms (HINKELMANN, 2005).

4.3.1 One-Step Methods

One-step methods determine the solution function on the new time level $n+1$ in just one step and generally take only the current time step n into account (an exception is the *Leap-Frog*-Method for example). The temporal derivative is approximated by forward differencing:

$$\frac{\partial e}{\partial t} \approx \frac{e^{n+1} - e^n}{\Delta t^n} \tag{4.7}$$

In *DiaTrans*, a *fully-implicit or Backward Euler Method* (HINKELMANN, 2005) is used, where e is determined on the new time level $n+1$, leading to eq. 4.8.

$$\frac{e^{n+1} - e^n}{\Delta t^n} = Ae^{n+1} \tag{4.8}$$

In a fully-implicit method, the unknowns of a center cell **C** on the new time level $n+1$ are determined using the unknown values of the neighboring cells (**W**, **E**, **N**, **S**) on the new time level and the known values of the center cell **C** on the current time level n.

Therefore, the unknowns on the new time level $n+1$ depend on each other, leading to a system of equations. The computational effort to solve this system is high, but in terms of stability reasons, there is no limit to the time step size. The order of consistency is

4.3 Discretization of Rate of Accumulation

only $O(\Delta t)$ (according to HINKELMANN, 2005), which has proven to be good enough for the applications carried out with *DiaTrans*.

Generally, e could also be computed between the current time level n and the new time level $n+1$. Then the *Crank-Nicholson Method* is obtained with the Crank-Nicholson factor θ with $0 \leq \theta \leq 1$ (for further information see HINKELMANN, 2005):

$$\frac{e^{n+1} - e^n}{\Delta t^n} = \theta A\, e^{n+1} + (1-\theta) A\, e^n \qquad (4.9)$$

For highly nonlinear systems of equations as occurring in *DiaTrans*, a combination of the *fully implicit* method and a *fully upwinding* technique for the discretization of the advection term (see sec. 4.4.2) has to be proven to yield the best results.

4.3.2 Implementation

For the multi-component formulation in *DiaTrans*, again e reads as $(\phi \rho_{mol} x^c)$. Assuming that the porosity ϕ does not change with respect to time, this leads to eq. 4.10 for a center cell **C** (see fig. 4.3) in the *FVM*:

$$\int_\Omega \frac{\partial e}{\partial t} = \int_\Omega [\phi \frac{\partial (\rho_{mol} x^c)}{\partial t}] dV \qquad (4.10)$$

Applying the product rule to eq. 4.10 yields:

$$\int_\Omega [\phi \frac{\partial (\rho_{mol} x^c)}{\partial t}] dV = \int_\Omega [\phi x^c \frac{\partial \rho_{mol}}{\partial t} + \phi \rho_{mol} \frac{\partial x^c}{\partial t}] dV \qquad (4.11)$$

When applying eq. 4.7 to eq. 4.11, the discretized form of the rate of accumulation term reads as the following:

$$\int_\Omega [\phi x^c \frac{\partial \rho_{mol}}{\partial t} + \phi \rho_{mol} \frac{\partial x^c}{\partial t}] dV \approx \frac{\phi \Delta x\, \Delta y\, \Delta z}{\Delta t^n} \{(x^c)^{n+1}[\rho_{mol}^{n+1} - \rho_{mol}^n] + \rho_{mol}^{n+1}[(x^c)^{n+1} - (x^c)^n]\} \qquad (4.12)$$

In *DiaTrans*, either a constant time step size or the adaptive time-stepping technique (sec. 4.12) can be utilized for Δt^n, in which the maximum feasible time size is always automatically obtained by just defining a minimum and a maximum time step size at the beginning of the simulation.

4 Numerical Methods & Implementation

4.4 Discretization of Advection

4.4.1 Implementation

For the FVM, the advective flux $\underline{J_a}$ (eq. 3.14) in eq. 3.63 is integrated across all four interfaces of the control volume. The general form is described in eq. 4.13:

$$\int_{\Gamma} \underline{J_a}\underline{n}dO = \int_{\Gamma} [-x^c \rho_{mol} \underline{\underline{\frac{k}{\mu}}} (grad\,p - \rho_{mass}\underline{g})]\underline{n}dO \qquad (4.13)$$

In the following, eq. 4.13 is applied to the east interface (**e**) and the south interface (**s**) of the control volume in the FVM to describe the process of discretization. Generally, it is assumed that the coordinate directions coincide with the main flow directions and that the porous medium is of isotropic nature. This leads to a description of the permeability tensor $\underline{\underline{k}}$ which reads as follows in a two-dimensional case:

$$\underline{\underline{k}} = \begin{bmatrix} k & 0 \\ 0 & k \end{bmatrix} \qquad (4.14)$$

Further, the gravitational vector $\underline{g}$ is described as shown in eq. 4.15 for the two-dimensional case:

$$\underline{g} = \begin{pmatrix} 0 \\ g \end{pmatrix} \qquad (4.15)$$

with g being equal to $-9.81\,m/s^2$.

Following these assumptions eq. 4.13 leads to eq. 4.16 and eq. 4.17 for the advective flux at the east and south interface (denoted as exponents **e** and **s** respectively):

$$\int_{\Gamma^e} [-x^c \rho_{mol} \underline{\underline{\frac{k}{\mu}}}(grad\,p - \rho_{mass}\underline{g})]\underline{n}dO^e \approx -(x^c \rho_{mol})^e \frac{k^e}{\mu}(grad\,p)^e(1)\Delta y \Delta z \qquad (4.16)$$

$$\int_{\Gamma^s} [-x^c \rho_{mol} \underline{\underline{\frac{k}{\mu}}}(grad\,p - \rho_{mass}\underline{g})]\underline{n}dO^s \approx -(x^c \rho_{mol})^s \frac{k^s}{\mu}(grad\,p - \rho^s_{mass}\underline{g})^s(-1)\Delta x \Delta z$$
$$(4.17)$$

In the two equations above the permeabilities k are averaged at the interface using the *harmonic* mean value (shown in eq. 4.18 for the east interface **e**), whereas the mass

4.4 Discretization of Advection

densities ρ_{mass} are averaged using the *arithmetic* mean value (shown in eq. 4.19 for the south interface **s**). The capital exponents correspond to the center cell (**C**), the eastern cell (**E**) and the southern cell (**S**) as outlined in fig. 4.3. Similar expressions can be obtained for all interfaces:

$$k^e = \frac{2\,k^C k^E}{k^C + k^E} \tag{4.18}$$

$$\rho^s_{mass} = \frac{\rho^C_{mass} + \rho^S_{mass}}{2} \tag{4.19}$$

Following these assumptions, the complete discretized equations are obtained for the east interface (eq. 4.20) and the south interface (eq. 4.21) from eq. 4.16 and eq. 4.17:

$$\int_{\Gamma^e} [-x^c \rho_{mol} \frac{\underline{\underline{k}}}{\mu}(grad\,p - \rho_{mass}\underline{g})]\underline{n}dO^e \approx -(x^c \rho_{mol})^e \frac{k^e}{\mu}(grad\,p)^e(1)\Delta y \Delta z =$$

$$-\frac{1}{\mu}(x^c \rho_{mol})^e \frac{2\,k^C k^E}{k^C + k^E}(\frac{p^E - p^C}{\Delta x})\Delta y \Delta z \tag{4.20}$$

$$\int_{\Gamma^s} [-x \rho^c_{mol} \frac{\underline{\underline{k}}}{\mu}(grad\,p - \rho_{mass}\underline{g})]\underline{n}dO^s \approx -(x^c \rho_{mol})^s \frac{k^s}{\mu}(grad\,p - \rho^s_{mass}\underline{g})^s(-1)\Delta x \Delta z =$$

$$\frac{1}{\mu}(x^c \rho_{mol})^s \frac{2\,k^C k^S}{k^C + k^S}(\frac{p^C - p^S}{\Delta y} - \frac{\rho^S_{mass} + \rho^C_{mass}}{2}g)\Delta x \Delta z \tag{4.21}$$

Similarly, the fully discretized equations can be obtained for the other two interfaces of the control volume (west and north respectively). Note, that the size of the normal vector is +1 for the east and north interface, but is -1 for the south and west interface. Eq. 3.70 can be used for the mass density ρ_{mass} to account for density-driven flow.

The term $(x\rho_{mol})$ in eq. 4.20 and eq. 4.21 can either be obtained arithmetically at the interface, which will yield the so-called *central method*, or a more sophisticated method like the *upwinding technique* can be chosen.

4.4.2 Upwinding Techniques

According to PATANKAR (1980), in a *fully upwinding technique* (also called upstream method) the value of an entity "ϕ at an interface is equal to the value of ϕ at the grid point [cell] on the upwind side of the face". This fully upwinding technique considers, that an advective dominated system is governed by the concentration of the upstream node and not by the one on the downstream side. Note that this method is stable, but shows a certain amount of numerical dispersion and is only of first order accuracy in space.

To be more flexible in handling more advective or diffusive controlled systems and to take care of high numerical dispersion, a modified upwinding technique is used in *DiaTrans* which allows to switch or to linearly interpolate between the upwind and the central method with an upwind parameter α ($0 \leq \alpha \leq 1$) to obtain the concentration ($x\rho_{mol}$) and therefore the advective flux at an interface. Note, that for practical applications and from experience the upwind parameter is usually chosen to be α ($0.5 \leq \alpha \leq 1$). If $\alpha = 1$, the fully upwinding scheme (upstream method) is obtained and $\alpha = 0.5$ leads to the central method.

The consequences for the discretized equations are only shown for the east and south interface. Again, similar expressions can be obtained for the other two interfaces.

Let v_e and v_s be the flow velocities in x-direction across the eastern interface (**e**) and in y-direction across the southern interface (**s**) calculated with the *Darcy* law (eq. 3.64). Then, the terms ($x\rho_{mol}$) in eq. 4.20 and eq. 4.21 can be substituted as shown in eq. 4.22 and eq. 4.23.

Note that the center cell **C** is the upstream cell for positive flow velocities in x-direction, whereas the neighbor cell **S** is the upstream cell for positive flow velocities in y-direction, following the definition of positive coordinate directions as defined in fig. 4.3.

$$(x^c \rho_{mol})^e = \begin{cases} \alpha[(x^c)^C \rho_{mol}^C] + (1-\alpha)[(x^c)^E \rho_{mol}^E] & if \ v_e \geq 0 \\ (1-\alpha)[(x^c)^C \rho_{mol}^C] + \alpha[(x^c)^E \rho_{mol}^E] & if \ v_e < 0 \end{cases} \quad (4.22)$$

$$(x^c \rho_{mol})^s = \begin{cases} \alpha[(x^c)^S \rho_{mol}^S] + (1-\alpha)[(x^c)^C \rho_{mol}^C] & if \ v_s \geq 0 \\ (1-\alpha)[(x^c)^S \rho_{mol}^S] + \alpha[(x^c)^C \rho_{mol}^C] & if \ v_s < 0 \end{cases} \quad (4.23)$$

The final discretization equations for the advection term for the east and south interface can now easily be obtained by combining eq. 4.20 and eq. 4.22 or eq. 4.21 and eq. 4.23, respectively.

Setting the upwind parameter α to 1.0, yielded the most stable results in *DiaTrans* and the combination of this *fully upwinding* technique and a *fully implicit* discretization in time leads also to the fact, that the solution is monotone at all times.

Generally, higher order methods can be employed in the *FVM*, such as the *Flux-Corrected Transport Methods (FCT)* or *Slope-* and *Flux-Limiter Methods*, which also preserve monotonicity. For a general overview and additional references about these methods, see HINKELMANN (2005).

4.5 Discretization of Diffusion & Dispersion

As for the advective flux, the diffusive/dispersive flux $\underline{J}_d$ (eq. 3.16) in eq. 3.63 is integrated across all four interfaces of the control volume. The general form is described in eq. 4.24:

$$\int_\Gamma \underline{J}_d \underline{n} dO = \int_\Gamma (-\phi \rho_{mol} \underline{\underline{D}}_{hyd} grad\, x^c) \underline{n} dO \tag{4.24}$$

Because of $\underline{\underline{D}}_{hyd}$ being a tensor and utilizing the definition of the hydrodynamic dispersion tensor, eq. 4.24 can also be written as shown in eq. 4.25:

$$\int_\Gamma \underline{J}_d \underline{n} dO = \int_\Gamma [-\phi \rho_{mol} \begin{pmatrix} D_{xx} + D_{m,e} & D_{xy} \\ D_{yx} & D_{yy} + D_{m,e} \end{pmatrix} \begin{pmatrix} \partial x^c/\partial x \\ \partial x^c/\partial y \end{pmatrix}] \underline{n} dO \tag{4.25}$$

In the following, eq. 4.25 is applied to the eastern interface (exponent **e**) of the control volume as an example for the discretization of $\underline{J}_d$ in the *FVM*.

In *DiaTrans*, a simplified approach for the hydrodynamic dispersion tensor is chosen, as generally in *DiaTrans* only the four direct neighbors of a center cell are regarded. In this approach, only the entries on the main diagonals are taken into account, neglecting the cross terms, i.e. $D_{xy} = D_{yx} = 0$. A lumped version of the dispersion tensor, in which the diagonal terms are set to $(D_{xx} + D_{m,e} + D_{xy})$ and $(D_{yy} + D_{m,e} + D_{yx})$ respectively and the off-diagonal terms are again set to 0, does not yield any significant differences. Eq. 4.25 then reads as shown in eq. 4.26 for the eastern interface:

$$\int_{\Gamma^e} \underline{J}_d \underline{n} dO^e = \int_{\Gamma^e} [-\phi \rho_{mol}(D_{xx} + D_{m,e}) \frac{\partial x^c}{\partial x}] \underline{n} dO^e \tag{4.26}$$

4 Numerical Methods & Implementation

Similar to the mass density for the advective flux, the molar density ρ_{mol} (eq. 4.27) and the porosity ϕ (eq. 4.28) are averaged arithmetically at the interface for discretization. Again, the exponents denote either the interface or the cell under consideration. Analogue expressions can be derived for the other three interfaces:

$$\rho_{mol}^e = \frac{\rho_{mol}^E + \rho_{mol}^C}{2} \qquad (4.27)$$

$$\phi^e = \frac{\phi^E + \phi^C}{2} \qquad (4.28)$$

Substituting the definition of D_{xx} from eq. 3.20, eq. 4.27 and eq. 4.28 into eq. 4.26, the final discretized equation for the diffusive / dispersive flux at the eastern interface can be obtained (eq. 4.29). For the other three interfaces, similar expressions can easily be obtained.

$$\int_{\Gamma^e} [-\phi \rho_{mol}(D_{xx} + D_{m,e}) \frac{\partial x^c}{\partial x}] \underline{n} dO^e \approx -\phi^e \rho_{mol}^e (D_{xx}^e + D_{m,e}) \frac{(x^c)^E - (x^c)^C}{\Delta x} (1) \Delta y \Delta z =$$

$$- \frac{\phi^E + \phi^C}{2} \frac{\rho_{mol}^E + \rho_{mol}^C}{2} (\alpha_l \frac{v_{ax}^2}{|\underline{v}_a^2|} + \alpha_t \frac{v_{ay}^2}{|\underline{v}_a^2|} + D_{m,e}) \frac{(x^c)^E - (x^c)^C}{\Delta x} \Delta y \Delta z \quad (4.29)$$

4.6 Discretization of Bioturbation

As for the diffusive / dispersive flux, the bioturbation flux $\underline{J}_{bt}$ (eq. 3.21) in eq. 3.63 is integrated across all four interfaces of the control volume. The general form is described in eq. 4.30:

$$\int_\Gamma \underline{J}_{bt} \underline{n} dO = \int_\Gamma (-\phi \rho_{mol} \underline{\underline{D}}_{bt} \, grad \, x^c) \underline{n} dO \qquad (4.30)$$

As the discretization for the bioturbation flux is nearly identical to the one of diffusion / dispersion, the implementation is simple due to the object-oriented nature of *DiaTrans*.

In the following, eq. 4.30 is applied to the northern interface (exponent **n**) of the control volume as an example for the discretization of $\underline{J}_{bt}$ in the FVM, which leads to eq. 4.31:

$$\int_{\Gamma^n} (-\phi \rho_{mol} D_{bt} \, grad \, x^c) \underline{n} dO^n \approx -\phi^n \rho_{mol}^n D_{bt}^n \, (grad \, x^c)^n (1) \Delta x \Delta z \qquad (4.31)$$

Again, the molar density and the porosity are averaged at the interface using the arithmetic mean as for the diffusive / dispersive flux. The bioturbation coefficient is averaged accordingly (eq. 4.32). The exponents denote either the interface or the cell under consideration and analogue expressions can be derived for the other three interfaces:

$$D_{bt}^n = \frac{D_{bt}^N + D_{bt}^C}{2} \quad (4.32)$$

Substituting the expressions for the arithmetic means into eq. 4.31 yields the final discretized equation for the bioturbation flux at the northern interface (eq. 4.33). The expressions for the other interfaces can be determined accordingly.

$$\int_{\Gamma^n} (-\phi \rho_{mol} D_{bt} \; grad \; x^c) \underline{n} dO^n \approx -\phi^n \rho_{mol}^n D_{bt}^n \; (grad \; x^c)^n (1) \Delta x \Delta z =$$

$$-\frac{\phi^N + \phi^C}{2} \frac{\rho_{mol}^N + \rho_{mol}^C}{2} \frac{D_{bt}^N + D_{bt}^C}{2} [\frac{(x^c)^N - (x^c)^C}{\Delta y}] \Delta x \Delta z \quad (4.33)$$

4.7 Discretization of Bioirrigation

The bioirrigation term B_i (eq. 3.24) in eq. 3.63 can be treated as a local source / sink term and thus it is discretized using the volume integral of a control volume in the FVM. Eq. 3.24 is therefore written in general form as shown in eq. 4.34:

$$\int_\Omega B_i dV = \int_\Omega [\alpha_{bi1} e^{(-\alpha_{bi2} b)} \phi \rho_{mol} (x_0^c - x^c)] dV \quad (4.34)$$

Applying eq. 4.34 to the control volume cell itself (i. e. **C** in fig. 4.3), eq. 4.35 is obtained.

$$\int_{\Omega^C} [\alpha_{bi1} e^{(-\alpha_{bi2} b)} \phi \rho_{mol} (x_0^c - x^c)] dV^C \approx \alpha_{bi1} e^{(-\alpha_{bi2} b)} \phi^C [\rho_{mol,0} x_0^c - \rho_{mol}^C (x^c)^C] \Delta x \Delta y \Delta z$$

$$(4.35)$$

The index 0 denotes a cell right at the sediment-water interface which yields the conditions of the porewater column. Note, that there is no need for averaging the porosity as only the center cell itself and no fluxes across interfaces are considered. Also the molar densities are not averaged to obtain $\rho_{mol}(x_0^c - x^c)$ in eq. 4.34, as the considered cells (one at the interface and the control volume cell itself) do not necessarily have a shared interface. Therefore, the mole fractions of those two cells [x_0^c and $(x^c)^C$] are multiplied with $\rho_{mol,0}$ and ρ_{mol}^C, respectively as shown in eq. 4.35.

4.8 Discretization of Reactions

As the reaction term r^c (eq. 3.36) in eq. 3.63 is treated as a local sink/source term term, it is discretized using the volume integral of a control volume in the FVM. The general form can be represented as in eq. 4.36:

$$\int_\Omega r^c \, dV \qquad (4.36)$$

Applying eq. 4.36 to the control volume cell itself (i.e. **C** in fig. 4.3), the discretized form (eq. 4.37) is obtained. Positive values of R_i denote accumulation of a component (i.e. $R_i > 0$) and negative values stand for consumption terms (i.e. $R_i < 0$).

$$\int_{\Omega^C} r^c \, dV^C \approx \phi^C [(\sum R_i)^c]^C \Delta x \, \Delta y \, \Delta z \qquad (4.37)$$

4.9 Discretization of Sinks & Sources

In *DiaTrans* local sink and source terms (q^c in eq. 3.63) can be defined for each component in certain areas of the computational domain using coordinate ranges for the horizontal x-direction and the vertical y-direction. Positive values of q^c denote sources (i.e. $q^c > 0$) and negative values stand for sink terms (i.e. $q^c < 0$).

Those terms use units of $mol/(m^3 \, s)$ as they are integrated over the control volume of a center cell.

As the sink/source term q^c is treated as a local term, it is discretized using the volume integral of a control volume in the FVM. The general form can be represented as in eq. 4.38:

$$\int_\Omega q^c \, dV \qquad (4.38)$$

Applying eq. 4.38 to the control volume cell itself (i. e. **C** in fig. 4.3), the discretized form (eq. 4.39) is obtained:

$$\int_{\Omega^C} q^c \, dV^C \approx (q^c)^C \, \Delta x \Delta y \Delta z \tag{4.39}$$

4.10 Initial and Boundary Conditions

According to HINKELMANN (2005), "for a unique solution of the initial boundary-value problem, initial conditions must be specified in the entire domain Ω as well as boundary conditions along the whole boundary Γ" for all primary variables.

In *DiaTrans*, initial (IC) and boundary conditions (BC) must be specified for each component separately.

4.10.1 Initial Conditions

The initial conditions describe the connection of the solution with the initial state of the solution. Following a formulation after HINKELMANN (2005), they are given in a form:

$$e(x, y, z, t = 0) = e_0(x, y, z) \tag{4.40}$$

where e_0 are the initial values of a scalar entity.

As the mole fractions are primary variables, the initial conditions are set for each component as mole fraction x^c over the whole computational domain defined by coordinate ranges in x- and y-direction. The mole fractions can easily be calculated from concentrations using eq. 3.6.

Instead of defining the mole fraction of the pure water component x^{H_2O} as initial condition, a pressure distribution of the total liquid phase as the other primary variable is defined. The mole fraction x^{H_2O} is calculated internally making use of the mass balance which states that the sum of all mole fractions is equal to 1 (eq. 3.62).

4.10.2 Boundary Conditions

After HINKELMANN (2005), "the boundary conditions, which may have different forms, represent the interaction with the surrounding domain." For *Dirichlet-boundary* conditions (denoted as e_D), which are also referred to as *Type I* boundary conditions, the solution function is described along the *Dirichlet* boundary (denoted as Γ_D) (see eq. 4.41):

$$e(x, y, z\, t) = e_D(x, y, z, t) \quad on \quad \Gamma_D \tag{4.41}$$

In *DiaTrans*, a *Dirichlet-boundary* condition corresponds to either a given pressure distribution along the boundary for the whole liquid water phase or given mole fraction distributions for all components except pure freshwater (H_2O). The *Dirichlet-boundary* conditions are defined using coordinates and coordinate ranges for the x- and y-direction.

When *Neumann-boundary* conditions (denoted as e_N), also named *Type II* boundary conditions, are used, HINKELMANN (2005) states that "the derivative of the solution function in the direction of the exterior normal vector must be given" along the *Neumann* boundary (denoted as Γ_N) (see eq. 4.42):

$$\frac{\partial e(x, y, z\, t)}{\partial \underline{n}} = e_N(x, y, z, t) \quad on \quad \Gamma_N \tag{4.42}$$

With respect to the general balance equations (eq. 3.58 and fig. 3.11), the *Neumann-boundary* conditions are defined as

$$\underline{F}_n = \underline{F} \cdot \underline{n} = \underline{F}(x, y, z, t) \cdot \underline{n} \quad on \quad \Gamma_N \tag{4.43}$$

with $\underline{F}_n$ describing a flux across the boundary Γ (HELMIG, 1997).

In *DiaTrans*, the *Neumann-boundary* conditions use units of $mol/(m^2\,s)$ as they are internally multiplied with the surface area of the inflow width, i.e. $\Delta x \Delta z$ for boundaries along the x-axis and $\Delta y \Delta z$ for boundaries along the y-axis respectively. If *no flow* boundaries are set, the *Neumann-boundary* value is set to 0. Inflow fluxes along the boundary are set using positive values for $\underline{F}_n$, whereas fluxes out of the computational domain are denoted using negative values for $\underline{F}_n$.

Also a linear combination of *Dirichlet-* and *Neumann-boundary* conditions is possible which would lead to *Cauchy-boundary* conditions (denoted as e_C), also called *Type III* boundary condition, along the *Cauchy* boundary (denoted as Γ_C). Although *Cauchy-boundary* conditions are not used in *DiaTrans*, their mathematical representation after

HINKELMANN (2005) is given in eq. 4.44:

$$\frac{\partial e(x, y, z\,t)}{\partial \underline{n}} + \alpha\, e(x, y, z\,t) = e_C(x, y, z, t) \quad on \quad \Gamma_C \tag{4.44}$$

Generally, it is possible in *DiaTrans* to use linear distributions of boundary condition values for the spatial coordinates, i. e. along a certain boundary. Also time-dependent boundary conditions can be set, using a generalized form to be able to account for any mathematical reasonable description using six values (see JABLONSKY, 2008). The first value describes the time t until which a certain definition of the boundary condition (defined by using either eq. 4.45 or eq. 4.46) is true:

$$e_D(t) = \alpha \cdot cos(\beta \cdot t + \gamma) + \delta + \epsilon \cdot t \tag{4.45}$$

$$e_N(t) = \alpha \cdot cos(\beta \cdot t + \gamma) + \delta + \epsilon \cdot t \tag{4.46}$$

In eq. 4.45 and eq. 4.46 the five parameters (α, β, γ, δ, ϵ) are used to mathematically describe the function of either $e_D(t)$ or $e_N(t)$.

4.11 Solving Nonlinear Systems of Equations

4.11.1 Introduction

The system of discretized balance equations, which are obtained from the partial differential equations is of mixed advection-diffusion-type, coupled, sparse, non-symmetric and highly nonlinear. Hence, the system has to be linearized before being solved. Thereced are different ways to linearize the equations, e. g. the *Picard* method or the *Newton-Raphson* method which is used in *DiaTrans* as it is a method of second order accuracy and shows better convergence behaviour compared to the *Picard* method, which is only a method of first order accuracy (HELMIG, 1997).

The principles of solving coupled and nonlinear systems of equations are outlined in fig. 4.4 and explained in the following. The nonlinear system of equations is solved using a nonlinear solver technique, e. g. the *Newton-Raphson* method, which is explained in detail in sec. 4.11.2. Inside the *Newton-Raphson* method the *Jacobi* matrix is computed leading to a coupled and linearized system of equations, which can now be solved using an iterative linear solver such as the *BiCGSTAB* method (see sec. 4.11.3) for non-symmetric systems of equations. As long as no convergence inside the nonlinear solver is achieved, the results of the linear solver are used to obtain updated starting values for the nonlinear solver. After convergence in the nonlinear solver is reached, the next time step inside a transient simulation can be computed. Details about the *Newton-Raphson* and the *BiCGSTAB* method are outlined in the following:

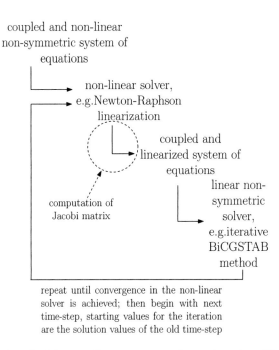

Figure 4.4: Principles of solving nonlinear systems of equations (modified after CLASS, 2004)

4.11.2 Newton-Raphson Method

The mathematical representation of the *Newton-Raphson* method which is used as nonlinear solver for the coupled nonlinear system of equations can be described according to PRESS et al. (2007) as follows.

When considering a problem with N functional relations F_i to be zeroed (i.e. the discretized balance equations), involving unknowns x_i with $i = 0, 1, ..., N-1$, eq. 4.47

4.11 Solving Nonlinear Systems of Equations

can be obtained:

$$F_i(x_0, x_1, ..., x_{N-1}) = 0 \quad i = 0, 1, ..., N-1 \tag{4.47}$$

We let $\underline{x}$ denote the entire vector of unknowns x_i and $\underline{F}$ denotes the entire vector of discretized balance equations (functions) F_i. According to PRESS et al. (2007), each of the functions F_i can be expanded in Taylor series in the neighborhood of $\underline{x}$ as shown in eq. 4.48:

$$F_i(\underline{x} + \delta \underline{x}) = F_i(\underline{x}) + \sum_{j=0}^{N-1} \frac{\partial F_i}{\partial x_j} \delta x_j + O(\delta \underline{x}^2) \tag{4.48}$$

The *Jacobian* matrix $\underline{J}$ is the matrix of partial derivatives appearing in eq. 4.48. The entries of the *Jacobian* matrix can be described as depicted in eq. 4.49:

$$J_{ij} = \frac{\partial F_i}{\partial x_j} \tag{4.49}$$

Eq. 4.50 shows eq. 4.48 written in matrix notation.

$$\underline{F}(\underline{x} + \delta \underline{x}) = \underline{F}(\underline{x}) + \underline{J} \cdot \delta \underline{x} + O(\delta \underline{x}^2) \tag{4.50}$$

Setting $\underline{F}(\underline{x} + \delta \underline{x}) = 0$ and neglecting terms of second order $O(\delta \underline{x}^2)$, a set of linear equations for the corrections (defect) $\delta \underline{x}$ can be obtained. The corrections $\delta \underline{x}$ move each function F_i closer to zero at the same time. Those equations can be described as shown in eq. 4.51:

$$\underline{J} \cdot \delta \underline{x} = -\underline{F} \tag{4.51}$$

Now, eq. 4.51 represents a coupled linear system of equations which can be solved directly using standard solver methods such as *BiCGSTAB* as it is done in *DiaTrans*.

Since often one does not know the analytical formulations of the partial derivatives of the various equations for the *Jacobian* matrix, or they may be too complicated to be calculated directly, FAIRES and BURDEN (1998) suggest to determine the *Jacobian* matrix entries using numerical differentiation by applying a finite difference approximation (eq. 4.52):

$$J_{ij} = \frac{\partial F_i}{\partial x_j} \approx \frac{F_i(..., x_{j-1}, x_j + \varepsilon, x_{j+1}, ...) - F_i(..., x_{j-1}, x_j, x_{j+1}, ...)}{\varepsilon} \tag{4.52}$$

4 Numerical Methods & Implementation

In eq. 4.52, ε is an infinitesimal small value, which controls the variation inside the finite difference approximation. Values such as $10^{-10} < \varepsilon < 10^{-6}$ have to be proven to be good estimates for ε. Generally, *DiaTrans* has shown to be very insensitive to the choice of ε.

After eq. 4.51 has been solved for the corrections $\delta\underline{x}$, those corrections are added to the solution vector of the unknowns $\underline{x}$ to obtain updated starting values for the next iteration step of the nonlinear solver (eq. 4.53). The whole process is iterated until convergence inside the nonlinear solver has been reached.

$$\underline{x}_{new} = \underline{x}_{old} + \delta\underline{x} \qquad (4.53)$$

The whole algorithm for the *Newton-Raphson* method as implemented in *DiaTrans* can be summarized as follows:

1. The initial conditions for the vector of unknowns $\underline{x}^0$ are used for the first time-step of the simulation.

2. The vector of functions $\underline{F}$ and the *Jacobian* matrix $\underline{\underline{J}}$ are calculated using $\underline{x}^0$.

3. The vector of corrections $\delta\underline{x}$ in eq. 4.51 is calculated with the inner linear *BiCGSTAB* method.

4. New starting values $\underline{x}_{new}$ for the nonlinear solver are obtained using eq. 4.53.

5. Steps 2–4 are repeated iteratively using $\underline{x}_{new}$ instead of $\underline{x}^0$ until convergence inside the *Newton-Raphson* method has been reached.

6. The next time-step $n+1$ is calculated using the values of $\underline{x}$ obtained at time-step n as initial conditions for the vector of unknowns $\underline{x}^0$ in step 1.

4.11.3 BiCGSTAB Method

As mentioned before, the *Newton-Raphson* method leads to a linear non-symmetric system of equations with a large number of unknowns which can be solved using a linear solver, which is the *BiCGSTAB* method in the case of *DiaTrans*. In general form, the system of equations can be written as:

$$AX = B \qquad (4.54)$$

with A being the coefficient matrix, X the vector of unknowns and B being the vector of the right-hand side. In case of eq. 4.51 obtained from the Newton-Raphson method, $A = \underline{\underline{J}}$, $X = \delta \underline{x}$ and $B = -\underline{F}$. As the coefficient matrix A is now independent of the vector of unknowns X, this system of equations is linear and can be solved with iterative linear methods such as *BiCGSTAB* (HINKELMANN, 2005). According to HINKELMANN (2005) who gives very detailed information about different solver techniques, "iterative methods operate with an iteration scheme: after an initial vector has been chosen, the solution is the limit of a series of approximations which is terminated if a stopping criterion which considers the approximation to be sufficient good is fulfilled." The initial guess is either the zero-vector or the result of the old time level for time-dependent problems. The stopping criterion can either be an absolute value or a relative value determined from the right-hand side vector. A pseudocode for the *BiCGSTAB* method is shown in fig. 4.5.

Conjugate Gradient methods (*CG* in *BICGSTAB*), for example, are very fast solvers for medium scale problems with about a few thousand to tens of thousands of unknowns. For non-symmetric matrices, the *BiConjugate Gradient Stabilized (BiCGSTAB)* method, which is a further development of some other methods, can be used in order to solve the system of linear equations as done in *DiaTrans*. According to HINKELMANN (2005), "the sequence of vectors which are orthogonal or conjugate to the matrix A is generated by a product of two polynomials (*Bi* in *BICGSTA*B), of which one locally minimizes the vector of the residuals and points in the direction of the steepest descent". A stabilized variant is used in the *BiCGSTAB* method, (*STAB* in *BICGSTAB*) which improves the convergence behaviour.

As a lot of sophisticated public-domain tool packages are available for iterative solvers as well as preconditioners (sec. 4.11.4), there is no need to code the solvers and preconditioners from scratch. In *DiaTrans*, the so-called *Matrix Toolkit for Java (MTJ)* (HEIMSUND, 2008) is used and parts of it, e. g. *BiCGSTAB* and *ILU*, are incorporated in the model structure which can easily be adjusted to the specific needs of a model. Additionally it is worth to mention that *MTJ* provides a large variety of linear solvers and preconditioners to be chosen from. The package is based on other toolkits, such as LAPACK – Linear Algebra PACKage (e. g. ANDERSON et al., 1999) and BLAS – Basic Linear Algebra Subprograms (e. g. BLACKFORD et al., 2002) and generally provides a "comprehensive collection of matrix data structures, linear solvers (direct and iterative), least squares methods, eigenvalue- and singular value decompositions" and "is

pseudocode BiCGSTAB Method

initial phase

$$r^{(0)} = B - AX^{(0)} \quad \text{for an initial guess } X^{(0)}$$
$$\text{chose} \quad \bar{r} \ (e.g.\ \bar{r} = r^{(0)})$$

iteration loop: k = 1, 2, ...

$$\lambda_{k-1} = \langle \bar{r}\, r^{(k-1)} \rangle$$
$$\text{if } \lambda_i = 0 : \text{stop}$$
$$\text{if } k = 1 : p^{(k)} = r^{(k-1)}$$
$$\text{if } k > 1 : \beta_{k-1} = (\lambda_{k-1}/\lambda_{k-2})(\delta_{k-1}/\Omega_{k-1})$$
$$p^{(k)} = r^{(k-1)} + \beta_{k-1}(p^{(k-1)} - \Omega_{k-1}\, v^{(k-1)})$$
$$\text{solve: } C\bar{p} = p^{(i)}$$
$$v^{(k)} = A\bar{p}$$
$$\delta_k = \lambda_{k-1}/\langle \bar{r}\, v^{(i)} \rangle$$
$$= r^{(i-1)} - \delta_k\, v^{(k)}$$
$$\text{solve: } C\bar{s} = s$$
$$t = A\bar{s}$$
$$\Omega_k = \langle t\, s \rangle / \langle t\, t \rangle$$
$$X^{(k)} = X^{(k-1)} + \delta_k \bar{p} + \Omega_k \bar{s}$$
$$r^{(k)} = s - \Omega_k\, t$$
$$\epsilon = \langle r^{(k)} r^{(k)} \rangle; \quad \text{if necessary continue}$$
$$\text{necessay for continuation: } \Omega_k \neq 0$$

storage: 2 matrices: A, C 10 vectors: $X, B, r, \bar{r}, p, \bar{p}, s, \bar{s}, v, t$

Figure 4.5: Pseudocode of the BICGSTAB Method (modified after LEPEINTRE, 1992)

designed to be used as a library for developing numerical applications, both for small and large scale computations" (HEIMSUND, 2008).

4.11.4 Preconditioning

In order to improve the convergence behaviour of the linear solver, a preconditioner can be used which pretreats the system of linear equations. According to HINKELMANN (2005), the overall aim during preconditioning is to reduce the number of iterations and the CPU time.

HINKELMANN (2005) also describes that preconditioning should be carried out in such a way that "certain properties of the system matrix are maintained, e. g. the symmetry." Also, the transferred system should have the same structure as the initial system, i. e. a sparse matrix structure should be obtained. Ideally, a preconditioner would yield the inverse of matrix A in eq. 4.54, but as this is costly in CPU time, the preconditioner only approximates the inverse of the initial matrix.

One of the most common and efficient preconditioning methods is the so-called *incomplete factorization*, in which the system matrix A is factorized by a lower matrix L and an upper matrix U (eq. 4.55):

$$A = LU \qquad (4.55)$$

An exact factorization would lead to a direct solution of the equations, but requires much CPU time and storage, because even in case of a sparse matrix A, L and U are dense (HINKELMANN, 2005). Therefore, incomplete factorizations can be used, in which only entries of the original structure are taken into account for the preconditioning step. The *ILU (Incomplete LU factorization)* method also belongs to the group of incomplete factorization methods and is used in *DiaTrans*. Generally, the convergence behaviour can be improved significantly using preconditioning and therefore CPU time can be saved. For further information about incomplete factorization see BRUSSINO and SONNAD (1989).

As mentioned before, the *Matrix Toolkit for Java (MTJ)* also provides preconditioners besides the iterative solvers, which can easily be implemented into the model structure. For *DiaTrans,* different preconditioning methods have been tested, with the ILU preconditioning yielding the best results in improving the convergence behaviour.

4.12 Adaptive Time-Stepping

4.12.1 Introduction

To enhance the CPU time of a simulation, an *adaptive time-stepping technique* is adopted in *DiaTrans*. After a minimum and a maximum time-step size have been defined by the user, this technique allows *DiaTrans* to obtain the maximum feasible time-step for a certain simulation during calculation.

This implementation makes use of the fact, that at the beginning of a simulation or during stages of big changes in concentrations, for example, small time-steps are needed, whereas at other times larger time-steps can be used without lacking accuracy. Therefore, the adaption of the time-step size must be possible in both directions, i. e. moving from small to bigger time-steps when it is physically possible during stages of few changes and moving from large to smaller time-steps when it is necessary during highly time-dependent phases of the simulation.

In the following, a detailed description of the implementation of this adaptive technique in *DiaTrans* is given. For more information, see JABLONSKY (2008).

4.12.2 Implementation

The *adaptive time-stepping* is implemented as follows in *DiaTrans*. The calculation procedure is started with the minimum time-step specified by the user in the Input-File Editor (see sec. 4.13.1). If the iterative process of the nonlinear *Newton-Raphson* method was successful, the time-step can be enlarged by a certain amount. It is also defined, that the time-step size is increased to an even bigger extent, when more than one successful nonlinear *Newton-Raphson* iterations have occurred subsequently without a failure in the iterative process. On the contrary, the whole *Newton-Raphson* iteration is repeated with a smaller time-step, when no convergence inside the nonlinear iterative process of solving the system of equations is obtained.

In order to numerically implement the behaviour described above, two factors are used to control the increase and decrease of the time-step size. The first factor ($fact1$), which is usually set to 1.2 at the beginning, is multiplied with the current time-step size (Δt) and hence influences its increase directly:

$$\Delta t^{new} = fact1 \cdot \Delta t^{old} \qquad (4.56)$$

The second factor ($fact2$), which is raised to the power of 1.1 after an successful *Newton-Raphson iteration*, is only multiplied with the first factor:

$$(fact1)^{new} = (fact2)^{1.1} \cdot (fact1)^{old} \qquad (4.57)$$

4.12 Adaptive Time-Stepping

Using this formulation, it is ensured that the time-step size is only increasing slowly at the beginning of a simulation, but faster after some simulation time has passed and when more than one successful nonlinear iterations have occurred subsequently.

Generally, the failure of the iterative process of the nonlinear solver can have different reasons. Either no distinct solution can be obtained from the inner linear solver (i.e. the *BiCGSTAB* method) or the outer nonlinear iterative solver inside the *Newton-Raphson* loop does not convergence or needs too many iterations. Inside *DiaTrans* it is checked for each time-step how many *Newton-Raphson* iterations are carried out in order to solve the system of equations. If the amount of iterations for the current time-step is far greater than for a previous time-step, it is defined that the nonlinear solver failed.

If an iteration of the nonlinear solver failed right after the time-step size has been increased, the same time-step is repeated with the last time-step size which could be successfully employed. If it fails again, the time-step size and the size of the first factor is halved (with the constraint that $fact1 \geq 1.0$). Also the second factor is decreased by extracting the square-root of it, which leads to a slower increase of the first factor for the following time-steps:

$$(fact2)^{new} = \sqrt{(fact2)^{old}} \qquad (4.58)$$

Applying this adaptive time-stepping technique to various problems, yielded big improvements in CPU time, especially in cases where no sharp concentration fronts occur during simulation, for example. Although this adaptation is not based on physical parameters, the time-step size adopts well to changing physical conditions inside the system. Note, that the choice of the minimum and maximum time-step size, which is specified by the user, is of overall importance as it controls the behaviour of a system at the beginning of a simulation (minimum time-step size) and determines how detailed certain processes and effects can be modeled during simulation without loosing important information (maximum time-step size).

4.13 Pre- & Postprocessing

4.13.1 Preprocessing

In order to specify a simulation, the *Input-File-Editor* is invoked in *DiaTrans*. All data which is necessary for a certain simulation, is defined in the editor, including information about the geometry, the components which shall be considered, the initial and boundary conditions, the reaction kinetics and the physical parameters such as permeability and porosity. As mentioned before, it is possible to use linear distributions and / or time-dependent information for the boundary conditions and the source / sink terms.

The *Input-File-Editor* is used interactively, as it checks during input if important information is missing or if the input format is wrong. Therefore, it is always ensured that a certain simulation is set up correctly. If necessary information is missing or wrong, the errors are highlighted inside the editor, so that the user is able to correct this information right away. Another useful feature inside the editor is, that hints for the input format pop-up, when the mouse is moved over the data input areas. This makes it a lot easier for users to set up a simulation flawlessly.

Also the output data, which should be analyzed during postprocessing, can be specified. If only some parameters are of interest for the user, this data can be specified in the editor. All other parameters and data is then neglected and not written to the output files which can save a lot data storage.

After all necessary data has been set, the simulation is started out of the editor. First the computational grid is built in the background with the information provided, then the calculation is carried out.

Fig. 4.6 shows a part of the Input-File-Editor of *DiaTrans* as an example.

4.13.2 Postprocessing

The visual representation of the simulation results is done with the open-source software package *Paraview* (KITWARE, 2008), which uses *.vtu-files as input. Any software package which can work with *.vtu-files (*.vtk-files respectively) could be used as standard postprocessing software, but *Paraview* provides lots of useful tools for data

4.13 Pre- & Postprocessing

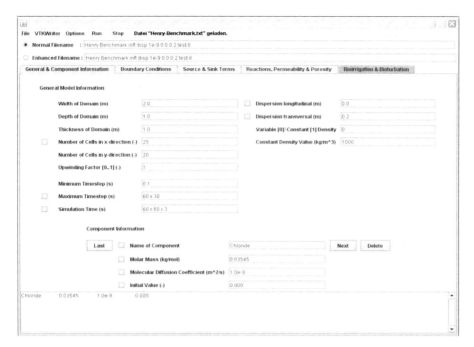

Figure 4.6: Input-File Editor of *DiaTrans*

analysis and is used as the standard postprocessing software for *DiaTrans* including analysis of time series, cut planes etc.

Also an easy comparison of different time series of simulation results is possible, even if the time-steps of the various simulations differ. In this case, *Paraview* automatically orders the simulations results at different time steps into a chronological order.

As mentioned before, a lot of tools (called *filters* in *Paraview*) are implemented in the standard configuration of the program, e.g. cut planes through the domain at any given point in time and space or the possibility of plotting simulation results for any variable at any given point over time. Also all visual formatting (e.g. legends, axis description etc.) is left up to the user to fit his needs.

For each time step in *DiaTrans* an XML-file is set up to fit the *Paraview* *.vtu-file format. For *DiaTrans*, this routine is implemented in the class VTKWriter.java in the fileIO-package.

The VTKWriter class is built to write the simulation results of each time step into an XML-file. The simulation results include values for pressure, mole fractions, *Darcy*

4 Numerical Methods & Implementation

and seepage velocities as well as the physical parameters such as permeability and porosity distributions. A "wrapper" or binding file is also created which combines all files of a transient simulation and information about the corresponding time values. It is therefore usable to present an animated view of the simulation results in *Paraview*.

Paraview utilizes many different file formats which can be processed. The one which is chosen in *DiaTrans* is called „UnstructuredGrid" and the corresponding suffix is *.vtu. Even *DiaTrans* only uses rectangular structured grids, it would be possible to use any geometry of the elements when using "UnstructuredGrid" as the standard file format.

All information, which is needed to create such an XML-file can be split into constant information and variable information, which can vary from one time step to another. Constant information does not change during a simulation run, such as the name of the simulation run, the list of all nodes and elements of the grid and the names of all components. This information is given to the class constructor and stays constant throughout the simulation run. Variable information can vary from one time step to another and includes information such as mole fractions of all components, pressure and velocity distributions.

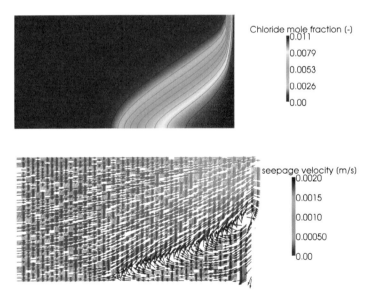

Figure 4.7: Exemplary output results of *Paraview* for the Henry problem

In fig. 4.7, exemplary output results of *Paraview* are shown for the *Henry* problem (see sec. 5.1), which was used as a standard verification run for density-driven flow and transport in the subsurface. This figure shows the simulation results for chloride mole fraction distribution with isolines (upper part) as well as seepage velocity vector distribution (lower part) as an example.

The actual time of a current time step is displayed in a specified format. The format used in *DiaTrans* separates the number of days, hours, minutes and seconds with a "0", so that it is easier to determine immediately at what point in time the output file was written. Therefore, it is not necessary anymore to convert the total number of seconds to obtain minutes, hours and days. This makes it a lot easier to follow a simulation vs. time.

5

Verification

To verify the reliability of *DiaTrans*, different benchmark tests are carried out. Various tests for density-driven flow and transport exist, with two of them applied to *DiaTrans*, i. e. the *Henry* problem for density-driven flow and transport only affected by advection and diffusion and the *salt dome* problem (also named HYDROCOIN) to model density-driven flow and transport including dispersion. The nature of those two benchmark tests and their results obtained with *DiaTrans* are presented in the first two parts of this chapter.

To verify the combination of transport, reactions, bioturbation and bioirrigation implemented in *DiaTrans*, standard diagenetic reaction processes in columns are used as a verification example. As no typical benchmark tests exist for this kind of simulation, major findings of BOUDREAU (1996), JOURABCHI et al. (2005) and WANG and VAN CAPPELLEN (1996) are chosen to verify the functionality of *DiaTrans*.

5.1 The Henry Problem

5.1.1 Introduction

HENRY (1964) presented a semi-analytical steady-state solution for a two-dimensional problem of groundwater flowing towards a seawater boundary. This leads to the advance of the saltwater front into a confined aquifer which was initially filled with uncontaminated freshwater. As a semi-analytical solution is available for the *Henry* problem, this problem has become one of the standard benchmark tests to evaluate the reliability of density-driven numerical flow and transport models.

5 Verification

According to DIERSCH and KOLDITZ (2005), a number of authors obtained similar results, although they used quite different approximation methods (e.g. PINDER and COOPER, 1970; SEGOL et al., 1975; VOSS and SOUZA, 1987 or OLDENBURG and PRUESS, 1995). To date, no numerical model has been able to reproduce exactly the semi-analytical results obtained by *Henry*.

SEGOL (1994) showed, however, that HENRY (1964) eliminated mathematical terms, which he thought to be insignificant, from the solution because of computational reasons. This led to a solution which was not exact. SEGOL (1994) then presented a revised calculation of the semi-analytical solution with the additional terms, which yielded slightly different results. SEGOL (1994) also showed, that numerical codes could now reproduce the correct answer of the *Henry* problem using the new solution. Note that SEGOL (1994) also used a different value for the molecular diffusion coefficient, as there have been some discrepancies in the use of this coefficient. A value of $D_m = 1.886 \cdot 10^{-5}\,m^2/s$ (instead of $6.6 \cdot 10^{-6}\,m^2/s$) is used in his revised solution as the "effective" diffusion coefficient is the product of the molecular diffusion coefficient and the porosity.

Besides this new semi-analytical solution, the numerical results of other models which simulated the revised solution, are used in this work to compare the results obtained with *DiaTrans* (see sec. 5.1.3). These results, which have been first reported by OSWALD et al. (1996), include solutions obtained with *FEFLOW* (DIERSCH, 1994), *MARCEAU* (OLTEAN et al., 1994), *SALTFLOW* (FRIND and MOLSON, 1994) and *UG* (BASTIAN et al., 1995).

Note that according to DIERSCH and KOLDITZ (2005), the *Henry* problem has some deficiencies when used as a benchmark test as "an unrealistic large amount of diffusion is introduced which results in a widely dispersed transition zone" which "makes the solution smooth and rather nonproblematic". DIERSCH and KOLDITZ (2005) suggest that additional benchmark tests should be used, e.g. one for situations with a narrow transition zone such as the *salt dome* problem (see sec. 5.2).

5.1.2 Model Setup

The idealized aquifer for the *Henry* problem is shown in fig. 5.1. The dimensions of the two-dimensional cross-sectional domain in the *FVM* are 2 m in x-direction, 1 m in y-direction and 1 m in z-direction.

5.1 The Henry Problem

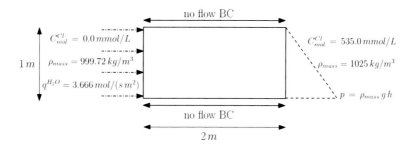

Figure 5.1: Model setup *Henry* problem

At the left-hand side boundary, a pure freshwater flux of $q^{Cl} = 3.666 \, mol/(m^2 \, s)$ using a *NEUMANN* BC is applied, which corresponds to $6.6 \cdot 10^{-2} \, kg/(m^2 \, s)$. The chloride concentration of the inflowing water at the left boundary is $C^{Cl}_{mol} = 0 \, mmol/L$ (*DIRICHLET* BC), corresponding to a mole fraction of chloride $x^{Cl} = 0.0$ and a mass density of $999.72 \, kg/m^3$ (see eq. 3.70 with $T = 10°C$).

At the right-hand side boundary hydrostatic pressure is assumed (*DIRICHLET* BC), with the seawater having a mass density of $1025 \, kg/m^3$. This leads to a maximum chloride concentration of $C^{Cl}_{mol} = 535 \, mmol/L$, i.e. maximum mole fraction of chloride $x^{Cl} = 0.00948175$.

Impermeable borders at the top and bottom exist, which are represented as no flow *NEUMANN* BCs for the two components pure water and chloride. As initial conditions, the whole domain is completely filled with pure freshwater, i.e. $x^{Cl}_{t=0} = 0.0$. All other relevant physical parameters are presented in table 5.1.

Table 5.1: Physical parameters for the *Henry* problem

Parameter	Value
molecular diffusion coefficient D_m	$1.886 \cdot 10^{-5} \, m^2/s$
longitudinal dispersion length α_l	$0.0 \, m$
transversal dispersion length α_t	$0.0 \, m$
permeability k	$1.02 \cdot 10^{-9} \, m^2$
porosity ϕ	0.35

The computational domain is discretized using 50 cells in x-direction and 50 cells in y-direction. Mesh convergence is reached. A *fully upwinding* scheme is used for the advection term (compare to sec. 4.4.2). The calculation is carried out using an unsteady

5 Verification

simulation, which leads to a steady-state situation after a certain amount of time. The maximum time-step size is 5 minutes and a steady-state situation is obtained after approximately 6 hours, although it can be stated that the situation after 128 minutes as shown in fig. 5.2 almost reflects the steady-state solution.

5.1.3 Results

Fig. 5.2 shows the chloride concentration distributions at different time-steps during the transient simulation of the *Henry* problem, whereas fig. 5.3 represents the *Darcy* velocity distribution of the steady-state solution.

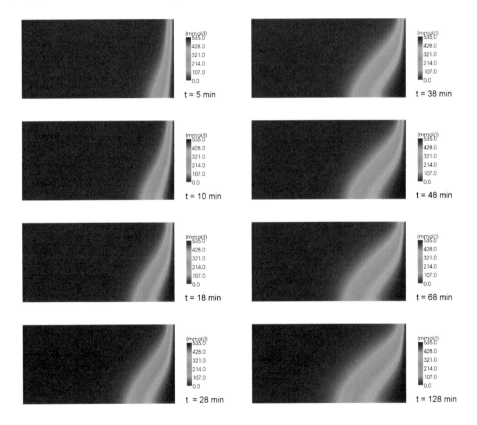

Figure 5.2: Chloride concentration distributions during the Henry problem

From fig. 5.2, the evolution of the saltwater front can clearly be depicted as it moves further into the freshwater aquifer as time evolves. After 128 minutes an almost steady-state solution exists with the moving saltwater front being in equilibrium with the inflowing freshwater from the left-hand side. The final numerical steady-state occurs after about 6 hours.

The *Darcy* velocity vectors in fig. 5.3 indicate that the saltwater is slowly swept into the domain from the lower part of the right-hand side boundary. The chloride-enriched water is then recirculated in the system and washed out of the system again across the right boundary. The highest velocities occur in the upper right-hand part of the domain as the total outflow of pure freshwater and mixed saltwater is bundled in this section of the aquifer. Note that the velocity vectors are all uni-length and the velocity magnitude is only characterized by different colors.

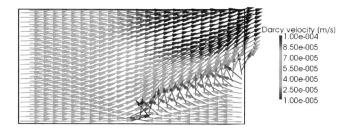

Figure 5.3: *Darcy* velocity distribution of the *Henry* problem

Fig. 5.4 shows a summary of model results for the 20%, 50% and 80%-isolines of maximum chloride concentration and the revised semi-analytical solution derived by SEGOL (1994). The results include solutions obtained with *FEFLOW* (DIERSCH, 1994), *MARCEAU* (OLTEAN et al., 1994), *SALTFLOW* (FRIND and MOLSON, 1994) and *UG* (BASTIAN et al., 1995), all of them first reported by OSWALD et al. (1996), and results of the *DiaTrans* simulation. OSWALD et al. (1996) reports that "the solutions obtained with these codes [*FEFLOW*, *MARCEAU*, *SALTFLOW* and *UG*] all match Segol's revised solution very closely, provided the appropriate diffusion coefficient $[1.886 \cdot 10^{-5}\, m^2/s]$ is used".

Also the results obtained with *DiaTrans* are in very good agreement with *Segol*'s revised solution. Only at the toes (i.e. at the bottom) of the isochlors minor differences are present, especially for the 20% and 50%-isochlors. Note that in *DiaTrans* the freshwater mass density is $999.72\, kg/m^3$ for $T = 10°C$ due to the density formulation used. The other numerical codes set the mass density to $1000\, kg/m^3$. Therefore, in *DiaTrans* the density difference is larger, which allows the saltwater front to move

5 Verification

further inland, i. e. leftwards into the computational domain. Overall, the results underline and confirm the strong capability of *DiaTrans* to simulate horizontally driven density-dependent flow and transport very reliable. Also, every numerical code introduces a certain amount of numerical diffusion. The shape of the isochlors simulated with *DiaTrans* suggest, that in *DiaTrans* a smaller amount of numerical diffusion is introduced compared to the other models, because for less diffusive solutions the saltwater front moves further inland as in the case of *DiaTrans*. An additional benchmark test for density-driven flow and transport which includes the effects of dispersion, the so-called *salt dome* problem, is presented in sec. 5.2.

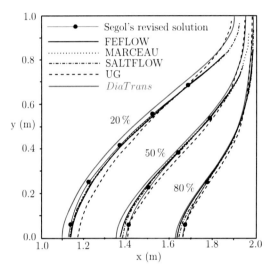

Figure 5.4: Result comparison *Henry* problem (modified after OSWALD et al., 1996)

5.2 The Salt Dome Problem

5.2.1 Introduction

As mentioned before, DIERSCH and KOLDITZ (2005) suggest that beside the *Henry* problem, additional benchmark tests should be used to evaluate the accuracy and reliability of density-dependent flow and transport models, especially for systems which are driven by vertical density differences. One, for situations with a narrow transition zone due to a low molecular diffusion coefficient, is the *salt dome* problem, which is also used herein as a benchmark test for *DiaTrans*.

The *salt dome* problem was proposed by participants of the international HYDROCOIN project for the verification of groundwater models (Swedish Nuclear Power Inspectorate 1986). Therefore, this benchmark test is also often referred to as HYDROCOIN problem. This test scenario uses a very simplified geometry to model "variable-density groundwater flow over a hypothetical salt dome" (DIERSCH and KOLDITZ, 2005). The assumed aquifer is 900 m in width and 300 m in depth and the salt dome is assumed to exist in the middle of the lower boundary, being 300 m wide. At the top boundary, a sloping pressure boundary is applied where the pressure decreases linearly and the chloride concentrations are zero. A detailed setup description is presented in sec. 5.2.2.

The *salt dome* problem has already been investigated by many authors, although no analytical solution exists (e. g. HERBERT et al., 1988; KOLDITZ et al., 1998; HOLZBECHER, 1998 or OLDENBURG and PRUESS, 1995). All results, which are obtained with different numerical codes, have been compared to each other and similar solutions were obtained. The simulation results of *DiaTrans* will also be compared to some of these results to confirm once again that its numerical code is reliable in modeling density-driven flow and transport.

Note that different test scenarios with varying values for the molecular diffusion coefficient and the dispersion lengths have been suggested by the HYDROCOIN group. To be able to compare the results of *DiaTrans* with other models, the test scenario from the study of HERBERT et al. (1988) is applied in this work, which is also used in the work of KOLDITZ et al. (1998) for both *FEFLOW* and *ROCKFLOW*.

From prior works mentioned above, a freshwater region with higher flow velocities is expected in the upper part of the domain, driven by the pressure difference on top. In the lower region, where the salt dome (brine pool) is present, the flow is expected to be recirculated with small velocities. The outflow of the water should be mainly focused in the upper right-hand corner of the domain.

5.2.2 Model Setup

The idealized aquifer for the *salt dome* problem is presented in fig. 5.5. The dimensions of the two-dimensional cross-sectional domain in the *FVM* are 900 m in x-direction, 300 m in y-direction and 1 m in z-direction.

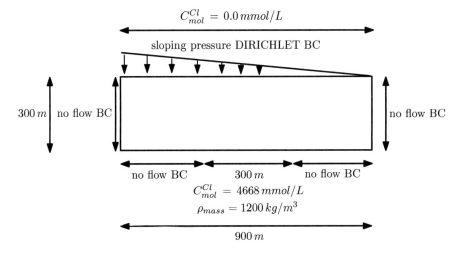

Figure 5.5: Model setup *salt dome* problem

At the left and right-hand side boundaries, no flow *NEUMANN* BCs for the two components pure freshwater and chloride are set. At the upper boundary, a sloping pressure DIRICHLET BC is applied where the chloride concentration is fixed to $C^{Cl}_{mol} = 0\,mmol/L$, corresponding to a mole fraction of chloride of $x^{Cl} = 0.0$. At the lower boundary, the left and right 300 m are also set as no flow NEUMANN BCs, whereas the middle 300 m represent the salt dome. The mass density of the brine in the dome equals $1200\,kg/m^3$, which corresponds to a maximum chloride concentration of $C^{Cl}_{mol} = 4668\,mmol/L$, i.e. mole fraction of chloride $x^{Cl} = 0.07512$.

As initial conditions, the whole domain is completely filled with pure freshwater ($x^{Cl}_{t=0} = 0.0$) with a mass density of $999.72\,kg/m^3$ at $T = 10°C$. All other relevant physical parameters are presented in table 5.1.

Table 5.2: Physical parameters for the *salt dome* problem

Parameter	Value
molecular diffusion coefficient D_m	$1.39 \cdot 10^{-8}\,m^2/s$
longitudinal dispersion length α_l	$20.0\,m$
transversal dispersion length α_t	$2.0\,m$
permeability k	$1 \cdot 10^{-12}\,m^2$
porosity ϕ	0.2
pressure difference on top Δp	$1 \cdot 10^5\,Pa$

The computational domain is discretized using 60 cells in x-direction and 30 cells in y-direction. Mesh convergence is reached. A *fully upwinding* scheme is applied again for the advection term (compare to sec. 4.4.2). Again, the calculation is carried out using an unsteady simulation, which leads to a steady-state situation after a certain amount of time. The maximum time-step size is 15 days and a steady-state situation is obtained after approximately 200 years.

5.2.3 Results

Fig. 5.6 and fig. 5.7 present results of the steady-state situation after about 200 years. From the chloride concentration distribution in fig. 5.6 it can clearly be seen, that the brine from the salt dome is swept into the system and transported towards the upper right-hand corner of the domain due to the flow induced by the sloping pressure boundary on top and diffusion / dispersion processes. Higher chloride concentrations represent the brine, i. e. enriched saltwater.

The distribution of *Darcy* flow velocities in fig. 5.7 confirm the expected flow pattern of higher flow velocities inside the freshwater region in the upper part of the domain and the outflow of the water mainly focused in the upper right-hand corner. In the lower region, where the influence of the salt dome is strong, streamlines (depicted as solid black lines) also indicate that the flow is recirculated with small velocities, as expected. In this region the *Darcy* velocities are so small that they cannot be recognized in the figure, as the length of the velocity vectors is scaled by the velocity magnitude for visualization.

5 Verification

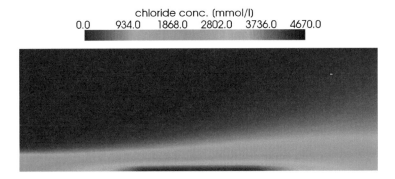

Figure 5.6: Chloride concentration distribution for the *salt dome* problem at steady-state

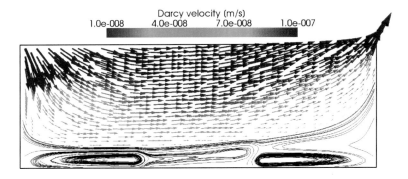

Figure 5.7: *Darcy* velocity distribution and streamlines for the *salt dome* problem

DIERSCH and KOLDITZ (2005) presented an overview of the results for the *salt dome* problem calculated by HERBERT et al. (1988) as well as results obtained with *FE-FLOW* and *ROCKFLOW*, both of them presented in the work of KOLDITZ et al. (1998). DIERSCH and KOLDITZ (2005) state that KOLDITZ et al. (1998) used different values for the molecular diffusion coefficient, but found "nearly identical salinity distributions".

Fig. 5.8 summarizes parts of the results mentioned above, including the simulation results of *DiaTrans*. In this figure, the 10 %, 20 % and 30 %-isochlors for the chloride concentrations at steady-state are outlined and compared to each other. Although no analytical solution exists for the salt dome problem, this comparison is useful to estimate the reliability of *DiaTrans* compared to other, well accepted, numerical codes.

5.2 The Salt Dome Problem

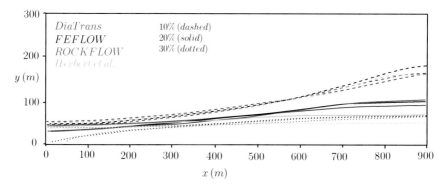

Figure 5.8: Result comparison for the *salt dome* problem (solutions from HERBERT et al., 1988; *FEFLOW* and *ROCKFLOW* (both after KOLDITZ et al., 1998) taken from DIERSCH and KOLDITZ, 2005)

When comparing the isochlors, almost all results are found to lie in a narrow range. Especially the comparison of *FEFLOW*, *ROCKFLOW* and *DiaTrans* yields good overall agreement between the models. Most notably, the isochlors of *FEFLOW* and *ROCKFLOW* are very similar, with the solution of *DiaTrans* fitting well into the results of the others. Only the results of HERBERT et al. (1988) differ from the other solutions, especially for the 20 % and 30 %-isochlors.

In summary, it can be stated that *DiaTrans* yields reliable results for the modeling of density-driven flow and transport also in the vertical direction. It is capable of reproducing the results of well accepted numerical codes, such as *FEFLOW* or *ROCKFLOW*, very accurately. The results of both the *Henry* and the *salt dome* problem show, that the numerical code inside *DiaTrans* is well developed and tested for density-driven flow and transport problems in subsurface sediments.

5.3 Standard Reaction Processes

5.3.1 Introduction

To verify the combination of the transport, reaction, bioturbation and bioirrigation terms which are implemented in *DiaTrans*, standard diagenetic reaction processes are simulated inside a simple column setup. As no typical benchmark tests exist for this kind of simulation, results of the works of BOUDREAU (1996), JOURABCHI et al. (2005) and WANG and VAN CAPPELLEN (1996) are used to qualitatively verify the reliability of *DiaTrans* for the processes mentioned above.

In this verification example, the main focus lies on the spatial concentration distributions in the uppermost part of the sediment column and the sequence of the main primary and secondary redox reaction processes (see sec. 3.9.1). All dissolved components mentioned in sec. 3.9 are included for the simulation, i.e. pure water, oxygen, nitrate, sulfate, sulfide, manganese, iron and methane. As a constant mass density is assumed, chloride is not modeled in this system. To investigate the primary and secondary reactions, all reactions defined in eq. 3.25 to eq. 3.35 are considered in the *DiaTrans* simulations. To investigate the influence of the bioturbation and bioirrigation terms, an additional simulation is carried out, which includes those terms, and the concentration profile results are compared to the standard setup.

Fig. 5.9 gives an overview of typical steady-state concentration profiles obtained by the authors mentioned above for the standard reaction processes. Note that all concentration profiles only represent the uppermost 10 cm in the sediment column.

Although different models and setups have been used by JOURABCHI et al. (2005), WANG and VAN CAPPELLEN (1996) and BOUDREAU (1996), the resulting concentration profiles, which can also be observed in the field in aquatic sediments, all show certain patterns which are also expected to be obtained with the *DiaTrans* simulations. Note, that for the *DiaTrans* setup, defined exponentially decreasing profiles for solid concentrations of organic matter (OM), manganese (MnO_2) and iron ($Fe(OH)_3$) are used (fig. 5.11) as only dissolved components are modeled inside *DiaTrans*.

The typical sequence of reaction processes can be identified from the profiles in fig. 5.9, e.g. a strong decrease of oxygen (O_2) concentrations within the uppermost few centimeters of the sediment column or the denitrification process (NO_3 decrease) which is only initiated when oxygen concentrations are low (fig. 5.9, top left and bottom left).

5.3 Standard Reaction Processes

Similarly, the reduction processes of solid manganese (MnO_2) and solid iron ($Fe(OH)_3$), which are the processes in which dissolved manganese (Mn^{2+}) and dissolved iron (Fe^{2+}) are formed, are only initiated when nitrate concentrations are low (fig. 5.9, bottom right). Also typical profiles for sulfate and sulfide concentrations are shown in fig. 5.9 (top right).

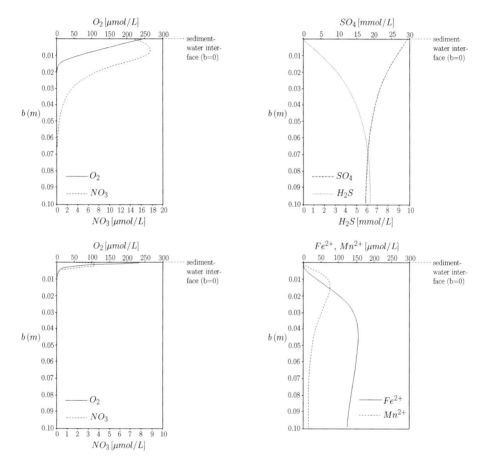

Figure 5.9: Typical concentration profiles for standard reaction processes (modified after JOURABCHI et al. (2005) – top left; BOUDREAU (1996) – top right & WANG and VAN CAPPELLEN (1996) – bottom left and right)

All these typical patterns should be obtained from the *DiaTrans* simulations to ensure the reliability of the reaction processes implemented in the model. The model run which includes bioturbation and bioirrigation should yield similar patterns, but with the

5 Verification

concentration profiles shifted further down into the sediment column, as higher oxygen, nitrate and sulfate concentrations are transported downwards due to bioirrigation. In the following, the model setup for the *DiaTrans* simulations is explained in detail, followed by the simulation results.

5.3.2 Model Setup

The idealized sediment column for the verification of standard reaction processes is presented in fig. 5.10. The dimensions of the two-dimensional cross-sectional domain in the *FVM* are 0.1 m in x-direction, 1 m in y-direction and 1 m in z-direction.

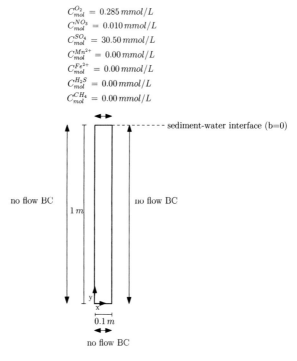

Figure 5.10: Model setup for standard reaction processes in columns

At the left and right-hand side boundaries and the lower boundary, no flow *NEUMANN* BCs are set for all components. At the upper boundary, a constant pressure DIRICHLET BC is applied for the liquid phase. Transport is only governed by diffusive processes as no advective transport – and therefore also no dispersive transport

5.3 Standard Reaction Processes

– is induced with this setup. This *quasi-one-dimensional* setup corresponds to the model setups of the other authors (i.e. BOUDREAU, 1996; JOURABCHI et al., 2005 and WANG and VAN CAPPELLEN, 1996). Assumed fixed molar concentrations, i.e. fixed mole fractions, are set at the upper boundary for all dissolved components, where $C_{mol}^{O_2} = 0.285\,mmol/L$, $C_{mol}^{NO_3} = 0.01\,mmol/L$, $C_{mol}^{SO_4} = 30.5\,mmol/L$ and $C_{mol}^{Mn^{2+}} = C_{mol}^{Fe^{2+}} = C_{mol}^{H_2S} = C_{mol}^{CH_4} = 0.0\,mmol/L$. The molecular diffusion coefficient is set to $D_m = 1.0 \cdot 10^{-9}\,m^2/s$ for all dissolved components, whereas the dispersion lengths are $\alpha_l = \alpha_t = 0.0\,m$.

As initial conditions, the whole domain is completely filled with pure freshwater, i.e. all molar concentrations of dissolved components are set to 0. The constant mass density of the liquid phase is assumed to be $1025\,kg/m^3$. All other relevant physical parameters are presented in table 5.3. The maximum concentrations at the upper boundary and the value for the porosity are chosen according to BOUDREAU (1996). For the first simulation, bioturbation and bioirrigation are set to 0, whereas the formulation and values for bioturbation (sec. 3.7 and eq. 3.23) and the formulation for bioirrigation (sec. 3.8) are used for the second model run. Using the bioirrigation parameters presented in table 5.3 leads to a small influence of the bioirrigation term only present in the uppermost part of the sediment column.

Table 5.3: Physical parameters for the standard reaction processes

Parameter	Value
permeability k	$1 \cdot 10^{-9}\,m^2$
porosity ϕ	0.8
bioirrigation parameters $\alpha_{bi1}/\alpha_{bi2}$	$10^{-7}\,\frac{1}{s}\,/\,50\,\frac{1}{m}$

A very fine mesh resolution in vertical y-direction is applied (400 cells), whereas only 3 cells in x-direction are used to save computational time as the concentration distributions are constant over the width of the domain. Again, the calculation is carried out using an unsteady simulation, which leads to a steady-state situation after a certain amount of time. The maximum time-step size is 1 hour and steady-state is reached after approximately 75 days.

The assumed exponentially decreasing profiles for the solid concentrations OM, MnO_2 and $Fe(OH)_3$, which are constant over time, are presented in fig. 5.11.

Shape and maximum values of those profiles are again similar to the steady-state distributions reported by BOUDREAU (1996); JOURABCHI et al. (2005) and WANG

5 Verification

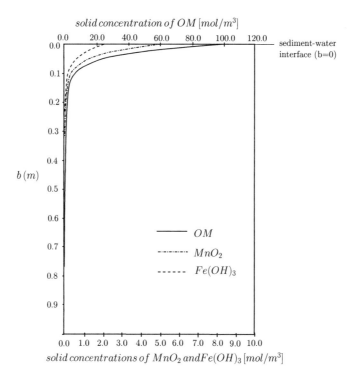

Figure 5.11: Solid concentration profiles as used for verification of standard reaction processes (expressed per unit volume total sediment.)

and VAN CAPPELLEN (1996). Certainly, the distributions of the solids, which are not simulated in *DiaTrans*, have a strong influence on the steady-state concentration distributions of the dissolved components as the solids govern the reaction kinetics of the primary reactions. The profiles used for the verification examples presented herein must be regarded as the best estimate available from literature. Note, that different shapes and values for the solid concentration profiles can be defined inside *DiaTrans* to adjust the model to other applications.

5.3.3 Results

The resulting steady-state concentration distributions for the standard reaction processes without bioturbation and bioirrigation are displayed in fig. 5.12 for dissolved oxygen, nitrate, sulfate, manganese, iron and sulfide.

– is induced with this setup. This *quasi-one-dimensional* setup corresponds to the model setups of the other authors (i. e. BOUDREAU, 1996; JOURABCHI et al., 2005 and WANG and VAN CAPPELLEN, 1996). Assumed fixed molar concentrations, i. e. fixed mole fractions, are set at the upper boundary for all dissolved components, where $C_{mol}^{O_2} = 0.285\,mmol/L$, $C_{mol}^{NO_3} = 0.01\,mmol/L$, $C_{mol}^{SO_4} = 30.5\,mmol/L$ and $C_{mol}^{Mn^{2+}} = C_{mol}^{Fe^{2+}} = C_{mol}^{H_2S} = C_{mol}^{CH_4} = 0.0\,mmol/L$. The molecular diffusion coefficient is set to $D_m = 1.0 \cdot 10^{-9}\,m^2/s$ for all dissolved components, whereas the dispersion lengths are $\alpha_l = \alpha_t = 0.0\,m$.

As initial conditions, the whole domain is completely filled with pure freshwater, i. e. all molar concentrations of dissolved components are set to 0. The constant mass density of the liquid phase is assumed to be $1025\,kg/m^3$. All other relevant physical parameters are presented in table 5.3. The maximum concentrations at the upper boundary and the value for the porosity are chosen according to BOUDREAU (1996). For the first simulation, bioturbation and bioirrigation are set to 0, whereas the formulation and values for bioturbation (sec. 3.7 and eq. 3.23) and the formulation for bioirrigation (sec. 3.8) are used for the second model run. Using the bioirrigation parameters presented in table 5.3 leads to a small influence of the bioirrigation term only present in the uppermost part of the sediment column.

Table 5.3: Physical parameters for the standard reaction processes

Parameter	Value
permeability k	$1 \cdot 10^{-9}\,m^2$
porosity ϕ	0.8
bioirrigation parameters $\alpha_{bi1}/\alpha_{bi2}$	$10^{-7}\frac{1}{s} / 50\frac{1}{m}$

A very fine mesh resolution in vertical y-direction is applied (400 cells), whereas only 3 cells in x-direction are used to save computational time as the concentration distributions are constant over the width of the domain. Again, the calculation is carried out using an unsteady simulation, which leads to a steady-state situation after a certain amount of time. The maximum time-step size is 1 hour and steady-state is reached after approximately 75 days.

The assumed exponentially decreasing profiles for the solid concentrations OM, MnO_2 and $Fe(OH)_3$, which are constant over time, are presented in fig. 5.11.

Shape and maximum values of those profiles are again similar to the steady-state distributions reported by BOUDREAU (1996); JOURABCHI et al. (2005) and WANG

5 Verification

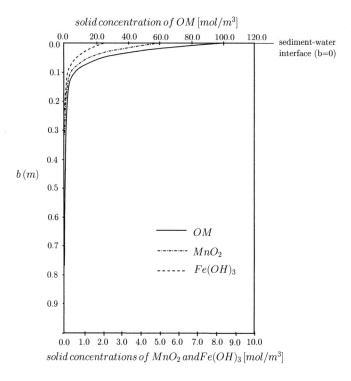

Figure 5.11: Solid concentration profiles as used for verification of standard reaction processes (expressed per unit volume total sediment.)

and VAN CAPPELLEN (1996). Certainly, the distributions of the solids, which are not simulated in *DiaTrans*, have a strong influence on the steady-state concentration distributions of the dissolved components as the solids govern the reaction kinetics of the primary reactions. The profiles used for the verification examples presented herein must be regarded as the best estimate available from literature. Note, that different shapes and values for the solid concentration profiles can be defined inside *DiaTrans* to adjust the model to other applications.

5.3.3 Results

The resulting steady-state concentration distributions for the standard reaction processes without bioturbation and bioirrigation are displayed in fig. 5.12 for dissolved oxygen, nitrate, sulfate, manganese, iron and sulfide.

5.3 Standard Reaction Processes

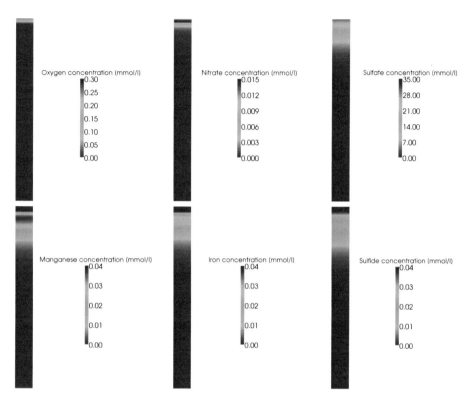

Figure 5.12: Concentration distribution for standard reaction processes (without bioturbation & bioirrigation)

It can clearly be seen, that all reaction processes take place only in the upper part of the sediment column, where high concentrations of organic matter (OM), solid manganese (MnO_2) and solid iron ($Fe(OH)_3$) are present. In greater depths these solid concentrations are small and therefore the driving forces for the primary redox reactions vanish (compare to fig. 5.11). Methane concentrations are negligible at steady-state and thus, they are not displayed. Similar results can be obtained for the situation with bioturbation and bioirrigation, with the only difference being the concentration distributions shifted to greater sediment depths.

From the concentration distributions in fig. 5.12, concentration profiles vs. depth are extracted for both situations (with or without bioturbation and bioirrigation). These results are summarized in fig. 5.13.

5 Verification

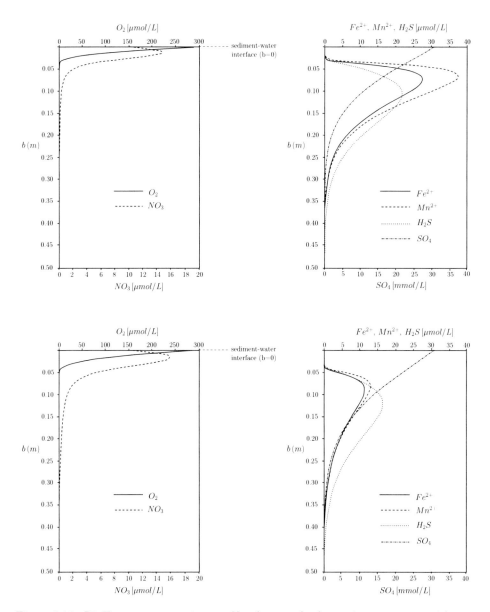

Figure 5.13: *DiaTrans* concentration profiles for standard reaction processes without (top) and with bioturbation & bioirrigation (bottom)

When analyzing the concentration profiles, some main features can be depicted. All reaction processes take place in the upper 45 cm of the sediment column. In the lower part, where the solid concentrations are 0, the component concentrations also turn out to be 0 at steady-state as all reaction rates approach 0. The concentration profiles obtained with *DiaTrans* (fig. 5.13) compare well to the typical steady concentration profiles reported by BOUDREAU (1996); JOURABCHI et al. (2005) and WANG and VAN CAPPELLEN (1996) (fig. 5.9).

As expected, the sequence of reaction processes can also be identified in the *DiaTrans* profiles. Note that the *DiaTrans* profiles describe the reaction processes in the upper 50 cm of the sediment, whereas the profiles in fig. 5.9 are only limited to the uppermost 10 cm. The strong decrease of oxygen concentrations within the uppermost few centimeters of the sediment column is evident, as well as the denitrification process (NO_3 decrease) at low values of oxygen concentrations (fig. 5.13, left). Likewise, the reduction processes of solid manganese (MnO_2) and solid iron ($Fe(OH)_3$), which are the processes in which dissolved manganese (Mn^{2+}) and dissolved iron (Fe^{2+}) are formed, are first occurring when nitrate concentrations are low (fig. 5.13, right). Profiles for sulfate and sulfide concentrations are also very similar to the ones reported by BOUDREAU (1996) (fig. 5.9, top right).

Overall, the agreement between the simulation results of *DiaTrans* and the results in fig. 5.9 is large and *DiaTrans* yields reliable results for the reaction kinetics, although the resulting concentration profiles can only be seen as qualitative examples as the model parameters may differ to the setups from the other authors. Moreover, *DiaTrans* only uses assorted reactions for the simulations in contrary to the authors mentioned above, who use a wider range of secondary redox reactions which will certainly also influence the results for the considered components to a certain extent. Differences may also be due to the fact, that the solid concentration profiles, which have a large impact on the primary redox reactions, are only assumed in *DiaTrans* based on values from literature.

The reliability of the bioturbation and bioirrigation terms implemented in the model can also be confirmed, when comparing the simulation results without these processes (fig. 5.13, upper part) and those, which include these effects (fig. 5.13, lower part).

The concentration profiles are clearly shifted further down into the sediment column as expected. Higher concentrations of oxygen, nitrate and sulfate are transported downwards from the overlying bottom water due to the effects of bioirrigation. The peak of high nitrate concentrations, for example, is wider because denitrification only

5 Verification

occurs deeper in the sediment column where oxygen concentrations become small. Similarly, solid manganese and solid iron reduction as well as sulfate reduction also set in at greater depths where oxygen and nitrate concentrations are small, following the typical successive pattern of the primary redox reactions. The profile for sulfide, which is formed when sulfate is reduced, is therefore also shifted to greater depths.

Due to the fact, that at greater depths the solid concentrations of OM, MnO_2 and $Fe(OH)_3$ are smaller, the maximum concentrations of dissolved manganese, dissolved iron and sulfide are clearly reduced for the case with bioturbation and bioirrigation. Although the single effect of bioturbation is small, as the maximum bioturbation coefficient is only $D_{bt} = 3.51 \cdot 10^{-10} m^2/s$ compared to the assumed molecular diffusion coefficient of $D_m = 1.0 \cdot 10^{-9} \, m^2/s$, the reliability of the implemented bioturbation and bioirrigation terms can clearly be confirmed during this verification process.

In summary, *DiaTrans* also yields reliable results for the modeling of primary and secondary reaction processes. It is highly capable of reproducing the results of other authors qualitatively. The results of this verification example show, that the numerical code inside *DiaTrans* is well developed for the above mentioned reaction processes.

6
Applications

This chapter presents different applications in the field of subsurface flow and transport modeling including reaction processes. The applications underline *DiaTrans'* large capability to simulate such processes.

First, the modeling of physical processes at *sand boils* in the Wadden Sea of Cuxhaven is investigated, including a sensitivity analysis, a comparison of field measurements and simulation results of chloride concentration profiles and the investigation of the tidal influence on maximum submarine groundwater discharge rates and on resulting concentration profiles. This preceding study has been performed using the numerical simulation software *MUFTE_UG* (e.g. HELMIG, 1997; HELMIG et al., 1998; HINKELMANN, 2005). The first part of this chapter closes with a comparison of *DiaTrans* and *MUFTE_UG*.

The modeling of biogeochemical reaction processes at submarine groundwater discharge sites in Eckernförde Bay, Baltic Sea, Germany is illustrated in the second part. Special focus lies on methane concentration profiles at so-called *vent* and *partial vent* sites, where submarine groundwater discharge rates as well as pore water methane concentrations have been extensively measured. Especially, the influence of bioirrigation on resulting methane concentration profiles in subsurface sediments is investigated and the simulation results are compared to existing field measurements.

The last part of this chapter deals with interactions of the processes occurring in the sediment and the overlying water column. This application highlights the need for further development of *DiaTrans* to better assess these interaction processes. To illustrate the main interests in this field, a rather simplified approach is presented which allows to qualitatively estimate the fate of contaminants or nutrients when entering the water column. This approach includes different simulations to evaluate the influence of

6 Applications

surface water velocities on the distribution of methane. Secondary reactions, such as methane re-oxidation by sulfate and oxygen, are also regarded in the water column as these reaction processes are not limited to the sediment and highly influence the fate of methane.

6.1 Modeling Physical Processes at Sand Boils

SCHANKAT et al. (2007) have presented an extensive study of the simulation of physical processes at *sand boils* to get a better understanding of and an enhanced insight into these processes. This preceding study includes a sensitivity analysis of the model and physical parameters as well as a comparison of model results and field measurements of chloride concentration profiles. An overview of this study, including general remarks on *MUFTE_UG*, which is the simulating software used for the study, is presented in the first three parts of this section.

The tidal influence on submarine groundwater discharges and chloride concentration profiles has been investigated in the work of SCHANKAT et al. (2009b). Special attention has been paid to maximum freshwater fluxes occurring during the tidal cycles of the overlying saltwater column. The main results of the work of SCHANKAT et al. (2009b) on the tidal influence at *sand boil* sites are summarized in the fourth part of this section, followed by a brief comparison of the models *MUFTE_UG* and *DiaTrans* for the modeling of submarine groundwater discharge at *sand boils*.

6.1.1 General Remarks on MUFTE_UG

As basis for the modeling approach used in the preceding study, the simulation tool *MUFTE_UG* (HELMIG, 1997; HELMIG et al., 1998; HINKELMANN, 2005) has been used. It represents a research code, programmed in *C*, with a strictly modular structure that allows the implementation of new physical model concepts together with new discretization techniques and solution methods along with, e.g. multigrid or parallelization strategies. *MUFTE_UG* was chosen for this first study of physical processes at *sand boils*, as it is a well-developed and accepted model in the field of subsurface flow and transport and it is also capable of simulating one-phase / two-component density-driven flow, although it has its strength in modeling multi-phase / multi-component flow and transport, for which it was designed in the first place. Because of this reason and the fact that the modular implementation of reaction processes is easier in an object-oriented framework and such a framework is extendable in the future in a simple manner, *DiaTrans* was developed for the direct coupling of flow, transport, reaction, bioturbation and bioirrigation processes in subsurface sediments.

Governing Equations

As mentioned in sec. 3.10.1, the model *MUFTE_UG* uses mass fractions X^c in contrast to *DiaTrans*, which defines the components as mole fractions x^c. In *MUFTE_UG*, the density-driven flow and transport processes are simulated by utilizing the two components freshwater (X^f) and saltwater (X^s), where the numerical modeling approach takes into account that freshwater and saltwater are fully miscible – the same concept which is employed in *DiaTrans*.

The continuity equation (eq. 6.1) for the flow and transport of the two components pure freshwater and pure saltwater in one fluid phase is similar to the one in *DiaTrans* (see eq. 3.61), neglecting the terms for reactions, bioturbation and bioirrigation. The *Darcy* law (eq. 3.64) is also used to replace the momentum equation for flow in porous media. Here, the sum of mass fractions adds up to 1 because of continuity reasons in contrast to *DiaTrans*, where the sum of mole fractions is equal to 1. The description of the hydrodynamic dispersion tensor follows the same approach as in *DiaTrans* (see eq. 3.15, 3.18, 3.19 and 3.20).

$$\phi \frac{\partial \rho_{mass} X^c}{\partial t} + div(X^c \rho_{mass} \underline{v}) - div\left(\phi \underline{D} \rho_{mass} grad\, X^c\right) - q^c = 0 \qquad (6.1)$$

According to OLDENBURG and PRUESS (1995), the mass density of the water mixture ρ_{mass} can be formulated as a function of the mass fractions and the densities of pure freshwater and pure saltwater (i.e. concentrated brine), respectively. This approach for the density-dependency as described in eq. 6.2 is used in *MUFTE_UG*.

$$\frac{1}{\rho_{mass}} = \frac{X^f}{\rho_{mass}^f} + \frac{X^s}{\rho_{mass}^s} \qquad (6.2)$$

6 Applications

Discretization and Solvers

For the preceding study, a two-dimensional model domain consisting of uniformly distributed quadrilaterals is used and the partial differential equations for the two components are discretized in space by a finite volume based collocation method, a so-called *box scheme* (e.g. HELMIG, 1997). For the flux terms a fully upwind technique is applied. The time is discretized with an implicit Euler scheme. Similar to the approach in *DiaTrans*, the equations lead to two coupled nonlinear partial differential equations for the components freshwater and saltwater. After discretization, a sparse system of algebraic equations exists for every time step, with possibly a large number of unknowns. As in *DiaTrans*, an outer *Newton-Raphson* iteration scheme (see sec. 4.11.2) is combined with an inner *BiCGSTAB* solver (see 4.11.3). For the solution, initial and boundary conditions are required for all primary variables, where the *NEUMANN* BCs are defined in $kg/(m^2 s)$ and the *DIRICHLET* BCs as pressure for the fluid phase or mass fraction for the component saltwater.

6.1.2 Sensitivity Analysis

In the following, the main results of the sensitivity analysis for the simulation of physical processes at *sand boils* carried out by SCHANKAT et al. (2007) are depicted. The aim of the study was to get a better understanding of and an enhanced insight into these processes.

Model Setup

The model domain used by SCHANKAT et al. (2007) is outlined in fig. 6.1. The domain represents the uppermost 100 cm of the coastal subsurface sediment in the Wadden Sea of Cuxhaven, Germany (see also sec. 2.1) which consists mainly of fine sand. The low-permeable lens in the upper part describes a peat layer which was found to be present during measurements at the site. The vertical column in the middle of the domain represents the higher permeable area of a *sand boil*, including the washed out part as described in fig. 3.2 (right). It is assumed, that in this column all the favoured flow paths of the upcoming freshwater can be found.

The model domain is 2.0 m wide in x-direction, 1.0 m high in y-direction and 1.0 m is assumed for the z-direction. Only for one simulation of the sensitivity analysis,

6.1 Modeling Physical Processes at Sand Boils

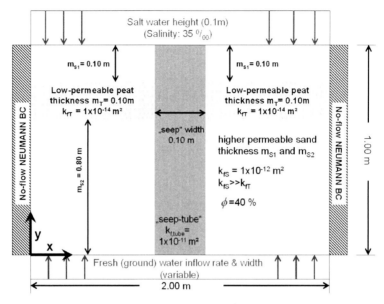

Figure 6.1: Model setup and boundary conditions (after SCHANKAT et al., 2007)

the domain width is changed to 4.0 m. All of the computations are performed on a rectangular grid with 1800 elements (60 x 30) for which mesh convergence is reached. The molecular diffusion coefficient D_m and the various values for permeability can be varied between different simulations.

At the left and right-hand boundaries no-flow *NEUMANN* BCs are set. At the lower boundary a constant flux *NEUMANN* BC is set with a variable freshwater inflow rate across variable inflow widths. At the inflow widths of pure freshwater, the mass fraction of saltwater is set to $X^s = 0$. At the upper boundary a constant head (i.e. constant pressure) *DIRICHLET* BC of pure saltwater ($X^s = 1$) is imposed with $p = \rho_{mass}\,g\,h$, where the height of the saltwater column h is fixed during all calculations ($h = 0.1\,m$). As initial condition, the domain is completely filled with pure saltwater with a hydrostatic pressure distribution. The model setup including boundary conditions and all relevant physical model parameters is displayed in fig. 6.1.

Although the flow through *sand boils* is actually of radial-symmetric nature, the 2D cross-sectional approach is chosen to get first insights into the physical processes occurring in the field. Radial-symmetric models, which *MUFTE_UG* and *DiaTrans* are not, or three-dimensional models would probably yield a more realistic representa-

6 Applications

tion of the processes, but the 2D cross-sectional approach is believed to represent the main processes accurately enough for the purpose of the studies of *sand boils* presented herein.

Chloride Concentration Profile Extraction

According to SCHANKAT et al. (2007), after each calculation with a different set of parameters, vertical profiles of the mass density ρ_{mass} of the water mixture are extracted at different locations – $x_0 = 0.10\,m$, $x_1 = 0.25\,m$, $x_2 = 0.50\,m$, $x_3 = 0.75\,m$ and $x_4 = 1.0\,m$. This mass density of the water mixture is calculated inside *MUFTE_UG* utilizing eq. 6.2. Then, eq. 3.66 and eq. 3.67 are used to convert the density profiles into chloride concentration profiles.

As a result, chloride concentration profiles such as shown in fig. 6.2 can be extracted where different graphs represent different qualitative profiles obtained at various locations. If a highly advective submarine groundwater discharge flux is present, an exponential decrease of chloride concentration can be observed over depth with the maximum chloride concentration right on top of the sediment (fig. 6.2, left). If the system is controlled mainly by diffusion / dispersion processes, the decrease is linear (fig. 6.2, middle) and if no submarine groundwater discharge influence is detected, the profiles stay constant on a high chloride concentration value over depth (fig. 6.2, right). This behaviour can also be observed in the field when measurements are taken using the sampling techniques described in sec. 2.2.

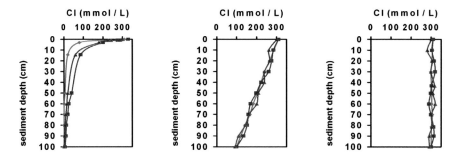

Figure 6.2: Chloride concentration profiles / advection (left) – diffusion/dispersion (middle) – no SGD influence (right) (after SCHANKAT et al., 2007)

For the intensive sensitivity analysis, a large number of simulations with different sets of parameters were carried out, all with the same submarine groundwater discharge rate at the bottom boundary of $1.0\,L/min$, an equal saltwater level of $0.10\,m$ at the top boundary with $\rho_s = 1025\,kg/m^3$, which corresponds to a salinity of $35\,\text{‰}$, a peat layer depth of $0.1\,m$ and a high permeable column width of $0.1\,m$. Permeabilities for the sand and peat areas are $10^{-12}\,m^2$ and $10^{-14}\,m^2$, respectively.

Sensitivity of Model Domain Parameters

For the sensitivity analysis, the model domain parameters such as domain width, submarine groundwater discharge inflow width and the permeability of the high permeable column are changed and the resulting chloride concentration profiles at locations x_0 to x_4 are compared. As a result, SCHANKAT et al. (2007) state, that only little differences can be seen at locations $x_0 = 1.0\,m$ to $x_2 = 0.50\,m$ (no differences at x_3 and x_4) when changing the model domain width from $2.0\,m$ to $4.0\,m$. The same applies for varying the permeability of the high permeable column between $10^{-10}\,m^2$ and $10^{-11}\,m^2$. Varying the SGD inflow width between $0.25\,m$ and $1.50\,m$ did not show any remarkable response at all of the locations $x_0 = 0.10\,m$ to $x_4 = 1.0\,m$. Exemplary, the results for location x_4 are shown in fig. 6.3 (left). SCHANKAT et al. (2007) state that "the sensitivity of all of the above mentioned model domain parameters can be neglected for the model setup used in this research [to model physical processes at *sand boils*]."

Sensitivity of Physical Parameters

Further, it is analyzed how sensitive the model responds to changes in D_m. Fig. 6.3 (right) shows the sensitivity of the molecular diffusion coefficient D_m on the resulting chloride concentration profiles at location $x_4 = 1.00\,m$ right in the centre of the *sand boil* obtained by SCHANKAT et al. (2007). Although the physically realistic molecular diffusion coefficient of chloride in pure water is about $1.0 \cdot 10^{-9}\,m^2/s$ according to SCHULZ (2000), it can be seen that the results react very sensitive when changing D_m between $6.6 \cdot 10^{-8}\,m^2/s$ and $6.6 \cdot 10^{-4}\,m^2/s$. For small values of D_m, the systems behaves more like an advective-controlled one at location x_4 right in the centre of the *sand boil*, for larger values of D_m more like a diffusion / dispersion controlled system (compare profiles of fig. 6.3 (right) to typical concentration profiles in fig. 6.2). The

6 Applications

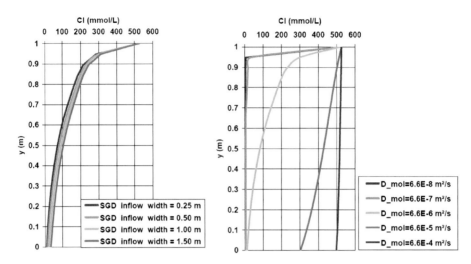

Figure 6.3: Left: Chloride concentration profiles at $x = 1.00\,m$ for different freshwater inflow widths; Right: Different molecular diffusion coefficients (both after SCHANKAT et al., 2007)

dispersion lengths α_l and α_t are set to $0.00\,\text{m}$ in this case. Similar results, i.e. high sensitivity, are obtained when varying the longitudinal and transversal dispersion lengths ($0.0001\,\text{m} < \alpha_l < 0.1\,\text{m}$ & $0.00001\,\text{m} < \alpha_t < 0.001\,\text{m}$) using a physically realistic and constant molecular diffusion coefficient of $D_m = 1.0 \cdot 10^{-9}\,m^2/s$. In that case, the study shows that an advective-controlled system is obtained for smaller dispersion lengths and a more diffusion / dispersion-controlled system for larger values of α_l and α_t.

6.1.3 Simulation vs. Nature

In the following, the main results of the comparison study by SCHANKAT et al. (2007) are outlined, in which the numerical modeling results are used to compare simulated chloride concentration profiles at *sand boils* with profiles obtained from field measurements. The measured profiles and the corresponding submarine groundwater discharge rates are gathered in the Wadden Sea of Cuxhaven, utilizing the measurement techniques described in sec. 2.2.

6.1 Modeling Physical Processes at Sand Boils

Model Setup

The model setup for the comparison of simulation results and field measurements which is employed in the work of SCHANKAT et al. (2007) is similar to the basic setup which is utilized for the sensitivity analysis (see fig. 6.1). This time, a *sand boil* with about 20 cm in width is chosen with an measured average submarine groundwater discharge rate of 0.5 L/min. The maximum chloride concentration in the overlying seawater is about $\sim 250\,mmol/L$, which corresponds to a saltwater mass density of $\sim 1012\,kg/m^3$. Therefore, the high permeable column width in the basic model setup is changed to 0.20 m, the saltwater density to $\rho^s{}_{mass} = 1012\,kg/m^3$, the lower *NEUMANN* BC to 0.5 L/min of SGD inflow with 1.00 m as inflow width. Permeabilities for the sand and peat areas are $10^{-12}\,m^2$ and $10^{-14}\,m^2$, respectively, whereas the permeability for the high permeable *sand boil* column is set to $10^{-11}\,m^2$. All other parameters are set as shown for the basic setup in fig. 6.1.

The molecular diffusion coefficient is assumed to be $D_m = 1.0 \cdot 10^{-9}\,m^2/s$, which corresponds to the value for chloride in pure water described in the literature (e.g. SCHULZ, 2000). The longitudinal dispersion length α_l and the transversal dispersion length α_t are the only parameters left to be variable for the comparison of simulation results and field measurements.

Results

Fig. 6.4 shows a typical distribution of the seepage velocity magnitude, which is in a range of $10^{-7}\,m/s$ to $10^{-5}\,m/s$. The velocity vectors, which are not scaled to the velocity magnitude in this case, clearly show that the modeled flow processes are highly two-dimensional and the 2D-modeling approach used in this work is essential to obtain reasonable results. The highest velocities occur right in the high permeable *sand boil* column with its favoured pathways as it was expected. Again, it has to be mentioned that the flow around *sand boils* is of radial-symmetric nature in reality. Therefore, a radial-symmetric or three-dimensional model approach might improve the results presented herein.

The resulting mass density distribution of the water phase is displayed in fig. 6.5 for the upper 50 cm of the marine sediment, with the maximum saltwater mass density on top and the minimum freshwater mass density at the bottom of the domain.

6 Applications

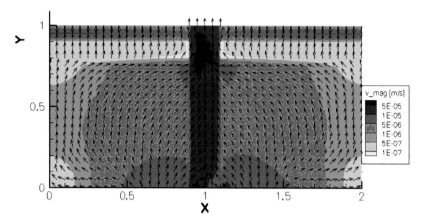

Figure 6.4: Seepage velocity distribution and vectors (after SCHANKAT et al., 2007)

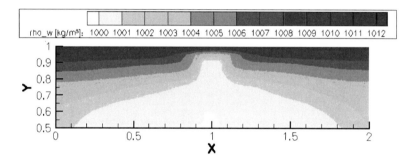

Figure 6.5: Water density distribution (after SCHANKAT et al., 2007)

Different dispersion lengths are used by SCHANKAT et al. (2007), always with the well established dispersion length ratio of $\alpha_l/\alpha_t = 10$ (e.g. HINKELMANN, 2005), to calibrate the model with the available measurements for chloride concentration profiles. Fig. 6.6 shows the measured and simulated chloride concentration profiles at locations $x_2 = 0.50\,m$ and $x_4 = 1.00\,m$. Measurements were only taken at three depths (i.e. 1 cm, 5 cm and 10 cm in the sediment) and compared to the simulated values. As the upper boundary fixes the saltwater density to its given maximum value, which may slightly differ in nature, the uppermost measurements at a depth of 1 cm are not used for comparison in the work of SCHANKAT et al. (2007). Simulated profiles at both locations x_2 and x_4 overestimate the chloride concentrations, when dispersion lengths of 1.00 mm (α_l) and 0.10 mm (α_t) are used (see also fig. 6.6).

6.1 Modeling Physical Processes at Sand Boils

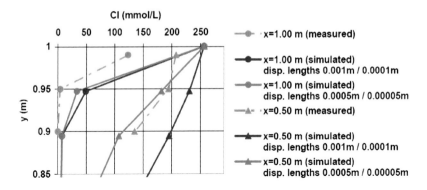

Figure 6.6: Chloride concentration profiles (after SCHANKAT et al., 2007)

Nevertheless, according to SCHANKAT et al. (2007), very good agreement between simulation results and measurements is achieved with dispersion lengths of 0.50 mm (α_l) and 0.05 mm (α_t) (see also fig. 6.6). These dispersion lengths are in the range of fine sand on the laboratory scale (HINKELMANN, 2005) and therefore reasonable for the scale of application. The effect of molecular diffusion is minor compared to the effects of dispersion when using dispersion lengths as chosen for this application. It should also be noted that dispersion lengths are calibration parameters and thus contain all other unknowns. Similar agreement can be obtained for other *sand boils* found in the Wadden Sea of Cuxhaven, i.e. for different submarine groundwater discharge rates or *sand boil* widths.

6.1.4 Tidal Influence

Based on the available field measurements as well as the density driven numerical transport model presented by SCHANKAT et al. (2007), SCHANKAT et al. (2009b) carried out a study with the specific objective of considering the impact of the tidal cycle on the flow field and the release / infiltration of freshwater or seawater, respectively. The results of this study are presented in the following.

Model Setup

The model setup (fig. 6.7) for the work of SCHANKAT et al. (2009b) is again similar to the one presented in sec. 6.1.2 and sec. 6.1.3, where the uppermost 100 cm of the coastal

6 Applications

subsurface sediment in the Wadden Sea of Cuxhaven is investigated. As by geological mapping (KURTZ, 2004) a stratification of permeable and less permeable sediment types has been observed with less permeable layers at 10 to 50 cm below the seafloor and very permeable sandy sediments at the base of these peat and clayey layers as well as at their top, the model domain depicted in fig. 6.7, which is again 2.0 m in width and 1.0 m in height, is chosen as the standard model setup.

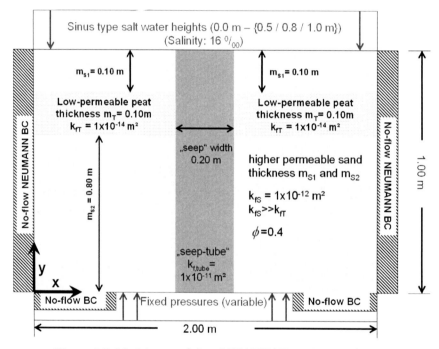

Figure 6.7: Model setup (after SCHANKAT et al., 2009b)

SCHANKAT et al. (2009b) performed all computations again on a rectangular grid with 1800 elements (60 x 30) and reached mesh convergence. On the left and right-hand boundaries no-flow *NEUMANN* BCs are set and at the lower boundary various constant pressure *DIRICHLET* BC are set for different model set-ups (i.e. 115000 Pa; 116500 Pa, 118000 Pa) from $x = 0.5\,m$ to $x = 1.5\,m$. Concentrations are fixed to 100 % of pure freshwater in this part. The remaining areas at the lower boundary are also set to no-flow *NEUMANN* BCs, as the previous study of SCHANKAT et al. (2007) showed that the upcoming freshwater is only present in a certain radius around the *sand boil* centre. Additionally, this previous steady-state study showed that a

6.1 Modeling Physical Processes at Sand Boils

pressure of 116500 Pa corresponds to a freshwater flow rate of approximately 0.5 L/min with a corresponding saltwater head of 0.1 m at the upper boundary. With the chosen pressures of 115000 Pa, 116500 Pa and 118000 Pa, which are just estimated in the study of SCHANKAT et al. (2009b) and have never been measured, it is expected to cover submarine groundwater discharge rates of approximately $0-2\,L/min$, as they have been observed in the field. A principle sketch of the model setup including boundary conditions and relevant physical parameters are outlined in fig. 6.7.

At the upper boundary, various sinus-type pressure *DIRICHLET* BCs are imposed for the different model setups, representing the tidal cycle variations of saltwater heights (0.0 m – {0.5; 0.8; 1.0 m}). Table 6.1 summarizes the relevant boundary condition parameters for the nine simulations, which have been carried out to obtain an improved insight into the influence of saltwater tidal cycles on pore water chloride concentrations and freshwater fluxes resulting from submarine groundwater discharge. As initial condition, the domain is assumed to be completely filled with pure saltwater with a hydrostatic pressure distribution.

Table 6.1: Boundary conditions for the various simulation setups (after SCHANKAT et al., 2009b)

Name	Saltwater heights at upper boundary	Pressure at lower boundary
MUFTE-AWI-20-01	0 – 0.5 m	116500 Pa
MUFTE-AWI-20-02	0 – 0.5 m	118000 Pa
MUFTE-AWI-20-03	0 – 0.5 m	115000 Pa
MUFTE-AWI-20-04	0 – 0.8 m	116500 Pa
MUFTE-AWI-20-05	0 – 0.8 m	118000 Pa
MUFTE-AWI-20-06	0 – 0.8 m	115000 Pa
MUFTE-AWI-20-07	0 – 1.0 m	116500 Pa
MUFTE-AWI-20-08	0 – 1.0 m	118000 Pa
MUFTE-AWI-20-09	0 – 1.0 m	115000 Pa

As in the simulation vs. nature example presented by SCHANKAT et al. (2007) (see sec. 6.1.3), a *sand boil* with a diameter of 20 cm is chosen in the work of SCHANKAT et al. (2009b). The maximum chloride concentration in the overlying seawater is again assumed to be $\sim 250\,mmol/L$, which corresponds to a saltwater mass density of $\sim 1012\,kg/m^3$. Permeabilities for the sand and peat areas are $10^{-12}\,m^2$ and $10^{-14}\,m^2$, whereas the permeability for the high permeable *sand boil* column is set to $10^{-11}\,m^2$. The molecular diffusion coefficient as well as the dispersion lengths correspond to

6 Applications

the values used by SCHANKAT et al. (2007) for the simulation vs. nature example ($D_m = 1.0 \cdot 10^{-9}\,m^2/s$; $\alpha_l = 0.5\,mm$; $\alpha_t = 0.05\,mm$).

Results

During the simulations of tidal cycles it can clearly be observed, that chloride concentration distributions and also flow velocity vectors change remarkably over time. During low tide, the upcoming freshwater leaves the system across the top boundary (indicated by velocity vectors in fig. 6.8).

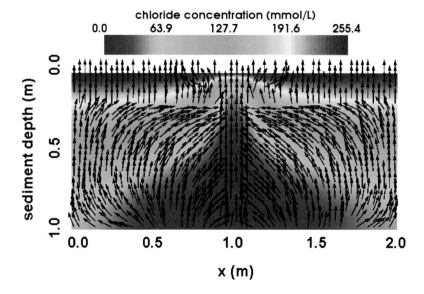

Figure 6.8: Chloride concentration distribution and velocity vectors during low tide (after SCHANKAT et al., 2008a)

If the pressure during high tide is big enough, a complete change in flow velocity directions can be observed, indicating that sea water enters the *sand boil* instead of freshwater from submarine groundwater discharge flowing out during low tide (fig. 6.9).

The velocity vectors are not scaled by the magnitude in this case and therefore only represent the direction of flow. Note, that both images (fig. 6.8 and 6.9) are presented in the work of SCHANKAT et al. (2008a) and have been obtained with *DiaTrans*. Similar figures could also be gained with *MUFTE_UG*.

6.1 Modeling Physical Processes at Sand Boils

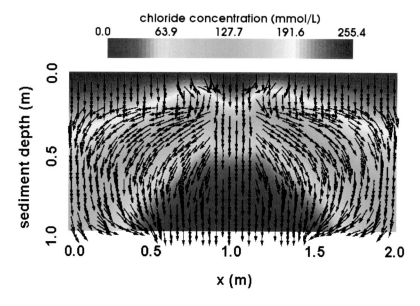

Figure 6.9: Chloride concentration distribution and velocity vectors during high tide (after SCHANKAT et al., 2008a)

Impact of Varying Saltwater Heights on Freshwater Fluxes

The impact of tidal cycles, represented by saltwater heights and implemented as the upper boundary condition, on freshwater fluxes Su_flux (SGD rates) for the time period of ~21 days to ~24 days after initialization is shown in fig. 6.10 for one of the nine simulation setups (i.e. MUFTE_AWI_20_06). The system reaches quasi-steady-state conditions after about 10 days. The values for the outflowing freshwater fluxes are obtained along the top boundary, where they represent submarine groundwater flowing through the system from bottom to top and entering the seawater column. Six tidal cycles are displayed in fig. 6.10, with two high tides and two low tides each day. It can clearly be seen, that the freshwater fluxes reach their minimum (i.e. Su_flux = 0.0 L/min) almost at the same time as the saltwater heights reach their maximum and vice versa. The small time shift of the sinusoidal graphs for saltwater heights and freshwater fluxes is due to the lagged physical response of the system caused by the density differences. The results are in quasi steady-state in this time period, i.e. the maximum values for Su_flux do not change from one tidal cycle to another and are

6 Applications

all in the range of ∼ 0.5 L/min. Similar results have been obtained for the other eight simulation setups (see table 6.1).

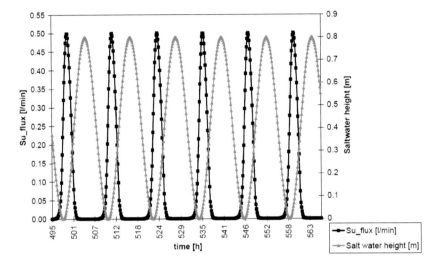

Figure 6.10: Freshwater flux (Su_flux) & saltwater height vs. time [MUFTE_AWI_20_06] (after SCHANKAT et al., 2009b)

Fig. 6.11 represents a smaller time period for the same simulation setup MUFTE_AWI _20_06. Following the arrows indicated in fig. 6.11 it can be shown, that the freshwater flux reaches the value of 0.0 L/min for this simulation set-up when the water level of the tide increases to a value of approximately 0.55 m. This value corresponds well with field measurements, which showed that at various *sand boils* the submarine groundwater discharge stopped when the tidal water level reaches ∼ 0.4 to 0.6 m. Again, similar results as shown in fig. 6.11 are obtained for the other eight simulation setups.

During the simulations, the change in total mass of freshwater (mass_Su) and saltwater (mass_Sa) present in the system are determined and are plotted together with the corresponding freshwater flux in fig. 6.12.

The total mass of freshwater plus saltwater stays constant at a certain point in time for each tidal cycle, representing again a quasi steady-state situation. According to SCHANKAT et al. (2009b), for increasing freshwater fluxes flowing through the system from bottom to top, an increase of total freshwater and a decrease of total saltwater in the system can clearly be noticed as expected. Once more, similar results are obtained for the other eight simulation setups.

6.1 Modeling Physical Processes at Sand Boils

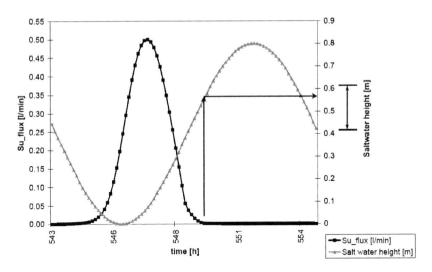

Figure 6.11: Estimation of corresponding saltwater height to zero freshwater flux (after SCHANKAT et al., 2009b)

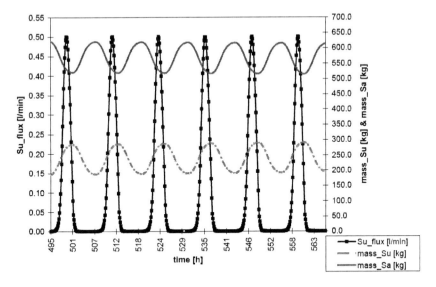

Figure 6.12: Freshwater flux (Su_flux), freshwater mass (mass_Su) & saltwater mass (mass_Sa) vs. time [MUFTE_AWI_20_06] (after SCHANKAT et al., 2009b)

6 Applications

Chloride Concentration Profiles During Tidal Cycles

After each simulation, chloride concentration profiles at the location $x_4 = 1.00\,m$ (centre of the *sand boil*) are extracted as described in sec. 6.1.2. Fig. 6.13 shows the results for three different simulation set-ups (MUFTE_AWI_20_04 to MUFTE_AWI_20_06) where the tidal cycle is always 0.00 – 0.80 m, but the pressure at the lower boundary differs (i.e. 115000 Pa, 116500 Pa and 118000 Pa, respectively).

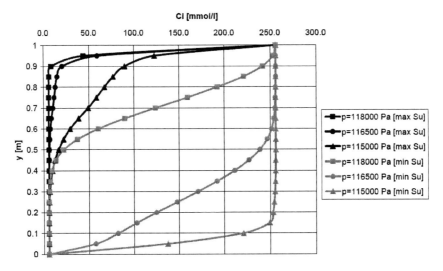

Figure 6.13: Chloride concentration profiles over depth at x= 1.0 m for different pressures at lower boundary and varying saltwater height (0.0 – 0.8 m) [MUFTE_AWI_20_04-06] (after SCHANKAT et al., 2009b)

For each of those three simulations, two profiles are shown – one where the freshwater flow through the domain is at its maximum value (max_Su), corresponding to low values of saltwater heights at the top boundary and one where it is at its minimum value (min_Su), i.e. 0.00 L/min, corresponding to a high value of saltwater heights.

When those profiles are compared to fig. 6.2, it can clearly be seen that the profiles for max_Su all show highly advection based shapes whereas the profiles for min_Su differ among each other. Looking at these graphs, a mixed advection / diffusion-dispersion based shape (p = 118000 Pa at the lower boundary), a diffusion / dispersion based shape (p = 116500 Pa) and a no-SGD-based shape (p = 115000 Pa) can be identified. From these findings it is evident that the shape of the chloride concentration profiles

6.1 Modeling Physical Processes at Sand Boils

is not only influenced by the saltwater height at the upper boundary but also by the pressure set for the lower boundary. Note, that all six profiles are fixed at the upper boundary to a maximum value of ~250 mmol/L of chloride concentration and a value of ~0 mmol/L at the lower boundary due to the numerical formulations used in this work.

Maximum Freshwater Fluxes

SCHANKAT et al. (2009b) present a summary of all maximum freshwater fluxes obtained from the nine different simulation setups during quasi-steady-state, i.e. MUFTE_AWI_20_01 to MUFTE_AWI_20_09 (see fig. 6.14).

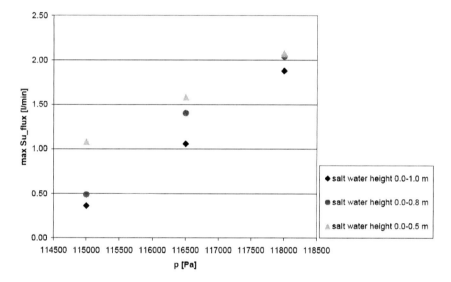

Figure 6.14: Maximum freshwater fluxes (after SCHANKAT et al., 2009b)

As the system does not physically react immediately to changes in boundary conditions due to the density-dependency of the water phase, the maximum freshwater fluxes are not the same for simulations with the same pressure at the lower boundary – even when the saltwater height is 0.0 m during a tidal cycle.

As expected, the maximum freshwater fluxes are in the range of $0-2\,\mathrm{L/min}$, which has also been observed in the field in the Wadden Sea of Cuxhaven. It can clearly be seen that higher pressures at the lower boundary lead to higher freshwater fluxes, whereas

6 Applications

higher tidal saltwater heights at the upper boundary result in lower freshwater fluxes. The maximum of 2.08 L/min is obtained with the set-up of MUFTE_AWI_20_02 and the minimum of 0.36 L/min results from MUFTE_AWI_20_09 (compare to table 6.1).

Although the flow through *sand boils* is more or less radial-symmetric in nature, all of the above mentioned results make it evident, that a two-dimensional cross-sectional model for density-driven flow and transport in the subsurface is highly capable of reflecting major findings, which are observed in the field during tidal cycles.

6.1.5 DiaTrans vs. MUFTE_UG

Even the benchmark tests for flow and transport (sec. 5.1 and 5.2) showed, that *DiaTrans* is capable of modeling these processes physically correct, *DiaTrans* is tested against results obtained with *MUFTE_UG* for a natural test case.

Therefore, the *MUFTE_UG* results for the chloride concentration profiles at a *sand boil* in the Wadden Sea of Cuxhaven are utilized to compare the two numerical models against each other.

Model Setup

The model setup for the comparison is identical to the one used for the *MUFTE_UG* simulation vs. nature example (sec. 6.1.3). The standard setup for this study is depicted in fig. 6.1 and the main standard parameters are as follows.

Again, the physical processes at a *sand boil* with about 20 cm in width are simulated with an measured average submarine groundwater discharge rate of 0.5 L/min, imposed at $x = 0.5 - 1.5\,m$ at the lower boundary as a *NEUMANN* BC. This leads to a value of $0.46296\,mol/(m^2\,s)$ in *DiaTrans*. As the *MUFTE_UG* results were obtained using a freshwater density of $1000\,kg/m^3$, the DIRICHLET BC for the chloride component is set to $x^{Cl} = 0.000105$.

The maximum chloride concentration in the overlying seawater is again assumed to be about $\sim 250\,mmol/L$, which corresponds to a saltwater mass density of $\sim 1012\,kg/m^3$. Therefore, a value of $x^{Cl} = 0.004606$ is used for the *DIRICHLET* BC in *DiaTrans*.

Permeabilities are set as in the simulation vs. nature example, i.e. $10^{-12}\,m^2$ and $10^{-14}\,m^2$ for the sand and peat areas, respectively and $10^{-11}\,m^2$ for the high permeable *sand boil* column in the middle of the domain. All other parameters are set as shown for the basic setup in fig. 6.1.

The molecular diffusion coefficient is always $D_m = 1.0 \cdot 10^{-9}\,m^2/s$, and for the first comparison of *DiaTrans* and the results of *MUFTE_UG*, the longitudinal dispersion length α_l and the transversal dispersion length α_t are set to the values which yielded the best results in *MUFTE_UG* (0.50 mm and 0.05 mm, respectively).

The *DiaTrans* results are obtained using a mesh resolution of 50 cells in simulated x-direction, 40 cells in simulated y-direction. The thickness of the domain in z-direction is set to 1.0 m. Mesh convergence is reached.

Results

The simulation results for chloride concentration profiles at locations $x_2 = 0.50\,m$ and $x_4 = 1.00\,m$ for both models are displayed in fig. 6.15. For the first comparison identical dispersion lengths have been used in both models. *DiaTrans* (red lines) underestimates chloride concentrations compared to *MUFTE_UG* (black lines) and therefore shows more advective-controlled based shapes of the profiles.

In a second simulation, larger dispersion lengths are used in *DiaTrans*, i. e. $\alpha_l = 1.0\,cm$ in longitudinal direction and $\alpha_t = 0.1\,cm$ in transversal direction. The value for the longitudinal dispersion lengths corresponds to fig. 3.6 after KINZELBACH (1992), assuming the length scale of the phenomenon being 1.0 m. Again, the well established dispersion length ratio of $\alpha_l/\alpha_t = 10$ (e. g. HINKELMANN, 2005) is applied to obtain the dispersion length value in transversal direction.

The results of this simulation are depicted as orange lines in fig. 6.15. It is obvious, that the *DiaTrans* results are now in close agreement with the MUFTE_UG concentration profiles. As already described in sec. 5.1.3, every numerical code introduces a certain amount of numerical diffusion. Again, the shape of the concentration profiles in fig. 6.15 simulated with *DiaTrans* suggest, that in *DiaTrans* a smaller amount of numerical diffusion is introduced compared to *MUFTE_UG* as the profiles reflect a more advection-controlled shape. It should also be noted again that dispersion lengths are calibration parameters and thus contain all other unknown processes of the underlying model concept.

6 Applications

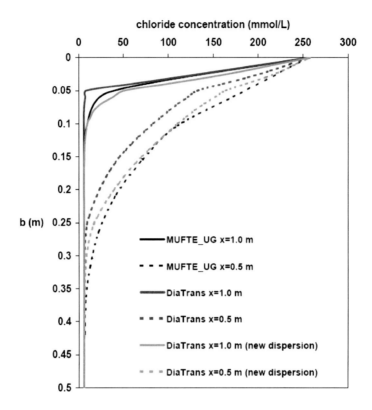

Figure 6.15: Chloride concentration profile comparison *MUFTE_UG* and *DiaTrans* at $x_4 = 1.00\,m$

No major differences in computational speed of the simulations could be determined between the two models carrying out the modeling with about the same number of elements (1800 in *MUFTE_UG* and 2000 in *DiaTrans*). The CPU time for the simulation was about 5 % lower in *MUFTE_UG* (developed in *C*) compared to the *Java* object-oriented approach implemented in DiaTrans. This underlines the findings of BULL et al. (2003), who shows that "the performance gap between Java and more traditional scientific programming languages is no longer a wide gulf" and that "the performance gap is small enough to be of little or no concern to programmers".

Overall, the results of this comparative study show that *DiaTrans* is capable of representing the density-driven flow and transport processes at *sand boils* occurring in nature, although its reliability could further be investigated with additional field data.

6.2 Modeling Reaction Processes at SGD Sites

This section deals with the modeling of biogeochemical reaction processes at submarine groundwater discharge sites, where the density-driven flow and transport is influenced by advective as well as diffusive / dispersive processes. This applications underlines the strength of *DiaTrans* to simulate highly coupled physical and biogeochemical processes in the subsurface, including all processes described in sec. 3.3 to sec. 3.9.

As an application example, the simulation of methane concentration pore water profiles at a study area in Eckernförde Bay, Baltic Sea, Germany is chosen, where so-called *vent* and *partial vent* sites have been identified and pore water measurements for dissolved methane concentrations are available. At these locations, submarine groundwater discharge has also been identified similarly to the *sand boil* sites in the Wadden Sea of Cuxhaven. Although SGD is not occurring in near-shore sediments in this case, but at the sediment-seawater-interface at greater water depths, the underlying processes and effects are identical to the ones at *sand boil* locations.

6.2.1 Introduction

HINKELMANN et al. (2002) carried out a study to simulate methane concentration distributions in subsurface sediments in Eckernförde Bay, Baltic Sea, Germany. Their work has been part of the joined EU *Sub-GATE* project, which major task has been to investigate SGD fluxes and transport from methane-rich coastal sedimentary environments. For more information on this project see SAUTER and SCHLÜTER (1999, 2000) and SAUTER (2001).

The study site in Eckernförde Bay is located about 25 km north-east of the town Kiel in the Baltic Sea, Germany (see fig. 6.16). The measured bathymetry of a typical *vent* location in the bay is also depicted in fig. 6.16. The lower left part of the figure indicates that sediment is flushed out at vent locations, whereas the lower right part verifies the findings, that a gas phase is measured in the sediment (HINKELMANN et al., 2002).

One task of the project has been the extensive continuous monitoring of fluid flow at *vent* sites (e. g. SAUTER and SCHLÜTER, 1999, 2000; SAUTER, 2001). Measurement results showed, that flow velocities from submarine groundwater into the seawater are relatively high and can be in the range of horizontal flow velocities in the bulk seawater.

6 Applications

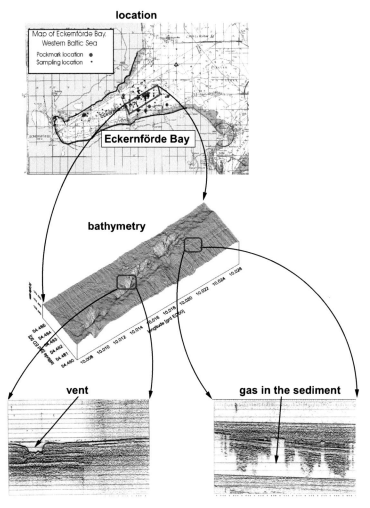

Figure 6.16: Location and bathymetry of a vent (after SAUTER, 2001)

Also pore water sampling has been carried out at so-called *vent* and *partial vent* locations, in order to describe the geochemical fluxes through the sediment-water-interface. Similarly to *sand boils*, *vents* do not only occasionally occur in Eckernförde Bay. Therefore, they certainly have a greater influence on the methane cycle in coastal environments than previously assumed. The measurements also showed, that methane formation and oxidation occurs within the soft mud sediments. When the maximum methane saturation in the liquid phase is exceeded, methane gas bubbles can appear, resulting in a gas phase within the sediment (e.g. HINKELMANN et al., 2000b). These multi-phase flow processes cannot be simulated with *DiaTrans*, as its model concept focuses on one-phase / multi-component flow and transport.

HINKELMANN et al. (2002) utilized a two-phase / three-component model to simulate the methane distributions in the sediment. As no corresponding data for methane formation and oxidation has been available, the methane formation has been regarded as a source term in the lower part of their domain with an estimated value. The oxidation processes of methane are simulated with an estimated sink term in the upper part of their domain. The maximum values for the source and sink terms have been described by biologists of the EU *Sub-GATE* project.

As *DiaTrans* is capable of modeling the biogeochemical reaction processes in the sediment, the approach utilized in this work is different. Similarly to HINKELMANN et al. (2002), high methane formation is assumed at the bottom of the sediment additional to the methane formation from organic matter described by the implemented *Monod* kinetics. In contrast to HINKELMANN et al. (2002), all solute components and their reactions described in sec. 3.9 are regarded. Therefore, methane oxidation is modeled using the implemented kinetics of primary and secondary reactions described in sec. 3.9.1, which is believed to yield more realistic results than utilizing a sink term, although the natural processes occurring in the sediment are even more complex.

6.2.2 Effects of Bioirrigation in Advective Flow

As bioirrigation (see sec. 3.8) is believed to be one of the predominant factors at locations less influenced by submarine fluid flow, a simple case study is carried out to estimate the effects of bioirrigation in advective flow. Two situations are simulated inside a sediment column, with one of them representing a situation without bioirrigation (simulation A) and the other one including this effect (simulation B). In both simulations, the effects of bioturbation (sec. 3.7) are considered. A submarine groundwater inflow discharge rate corresponding to a *partial vent* site in Eckernförde Bay, where the influence of fluid flow is less than at a *vent* location, is assumed at the lower boundary.

6 Applications

Model Setup

The idealized sediment column for the study of the effects of bioirrigation in advective flow is similar to the one used in the verification section (see sec. 5.3) and is presented in fig. 6.17. The dimensions of the two-dimensional cross-sectional domain in the FVM are again $0.1\,m$ in simulated x-direction, $1\,m$ in simulated y-direction and $1\,m$ in z-direction.

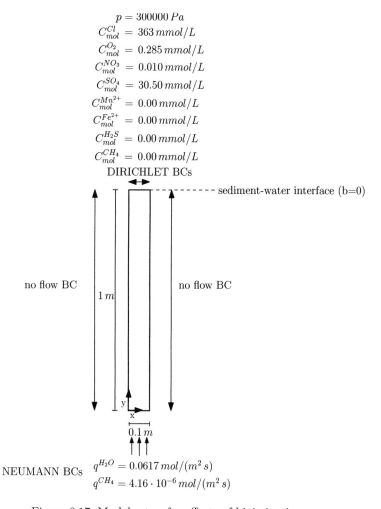

Figure 6.17: Model setup for effects of bioirrigation

At the left and right-hand side boundaries, no flow *NEUMANN* BCs are set for all components. At the upper boundary, a constant pressure DIRICHLET BC of $p = 300000\,Pa$ is imposed for the liquid phase, representing a seawater column of approximately 20 m. Assumed molar concentrations, i.e. fixed mole fractions, are set at the upper boundary for all dissolved components ($C_{mol}^{O_2} = 0.285\,mmol/L$, $C_{mol}^{Cl} = 363\,mmol/L$, $C_{mol}^{NO_3} = 0.01\,mmol/L$, $C_{mol}^{SO_4} = 30.5\,mmol/L$ and $C_{mol}^{Mn^{2+}} = C_{mol}^{Fe^{2+}} = C_{mol}^{H_2S} = C_{mol}^{CH_4} = 0.0\,mmol/L$). All molar concentrations, except the one for chloride, are assumed to be the same as used for verification (sec. 5.3). The molecular diffusion coefficient is set to $D_m = 1.0 \cdot 10^{-9}\,m^2/s$ for all dissolved components, whereas the dispersion lengths are assumed to be $\alpha_l = 0.01\,m$ and $\alpha_t = 0.001\,m$.

At the lower boundary, *NEUMANN* BCs are set for the two components freshwater (H_2O) and methane (CH_4) to induce an upward directed advective flow. The value for the freshwater component of $q^{H_2O} = 0.0617\,mol/(m^2\,s)$, corresponding to $4.0\,L/(m^2\,h)$, has been used by HINKELMANN et al. (2002) for a *partial vent* location in Eckernförde Bay and is utilized for this study. This value also corresponds to findings of KARPEN et al. (2006), who measured submarine groundwater discharge rates of $\sim 3 - 43\,L/(m^2\,h)$ in Eckernförde Bay at different locations.

As mentioned before, high methane formation is assumed at the bottom of the sediment additional to the methane formation from organic matter described by *Monod* kinetics. Therefore, $q^{CH_4} = 4.16 \cdot 10^{-6}\,mol/(m^2\,s)$ is set for the methane component, which is also reported by HINKELMANN et al. (2002), who used this as a maximum value for the source term in the lower part of their domain. The assumed exponentially decreasing profiles for the solid concentrations OM, MnO_2 and $Fe(OH)_3$, which are constant over time, are identical to the ones used during verification (sec. 5.3) and are presented in fig. 5.11.

As initial conditions, the whole domain is completely filled with saltwater with $\rho_{mass} = 1017\,kg/m^3$, corresponding to $C_{mol}^{Cl} = 363\,mmol/L$. The initial pressure distribution is hydrostatic and all other mole fractions of dissolved components are set to 0. Additional relevant physical parameters are presented in table 6.2.

For simulation A, only the effect of bioturbation (sec. 3.7) is taken into account, neglecting bioirrigation (sec. 3.8). For simulation B, also a large influence of bioirrigation is regarded with the values described in table 6.2, as the *partial vent* location is believed to be highly influenced by bioirrigation processes in this example. The transport in the sediment column is governed by density-influenced advective and diffusive / dispersive processes.

Methane re-oxidation by O_2 and SO_4 occurs in the uppermost part of the sediment column, where high concentrations of oxygen and sulfate are present. These re-oxidation reaction processes are modeled using the kinetics for secondary reactions as described in sec. 3.9.1.

Table 6.2: Physical parameters for effects of bioirrigation

Parameter	Value
permeability k	$1 \cdot 10^{-11}\,m^2$
porosity ϕ	0.9
bioirrigation parameters $\alpha_{bi1}/\alpha_{bi2}$	$10^{-5}\,\frac{1}{s} / 10\,\frac{1}{m}$

A fine mesh resolution in vertical y-direction is applied (50 cells), whereas only 3 cells in x-direction are used to save computational time as the concentration distributions are constant over the width of the domain. The maximum time-step size is 30 minutes and steady-state is reached after approximately 17 days for this quasi-1D simulation.

Results

Resulting methane and sulfate steady-state concentration distributions for simulations A and B are displayed in fig. 6.18. Note, that for both simulations the sediment column is initially assumed to be free of sulfate. Diffusive / dispersive processes allow sulfate to move downwards in the domain in direction of decreasing concentration gradients, whereas advection moves sulfate upwards and additionally transports methane to upper regions, where re-oxidation processes of methane with sulfate occur.

Fig. 6.18 (top right) underlines, that sulfate is only transported to a small amount by diffusion and dispersion for simulation A without the effects of bioirrigation. In contrast, high methane concentrations are present in the uppermost part of the column which are transported mainly by advection, fig. 6.18 (top left). The highly effective re-oxidation area is therefore limited to the upper 10 % of the domain. Although methane concentrations are zero at the top boundary due to the assumed *DIRICHLET* BC, the maximum methane concentration will reach the upper boundary and enter the seawater column in reality.

6.2 Modeling Reaction Processes at SGD Sites

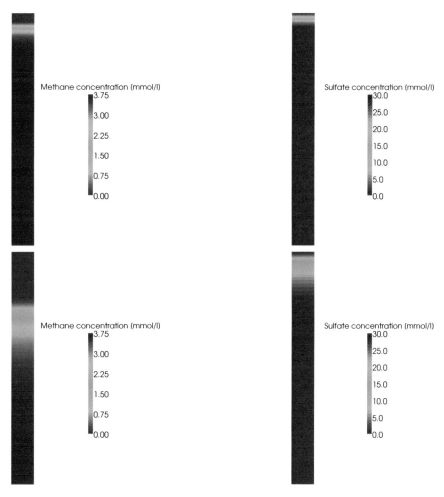

Figure 6.18: Methane and sulfate concentration distributions for simulation A without bioirrigation (top) and for simulation B with bioirrigation (bottom)

The effects of bioirrigation can clearly be seen for simulation B in fig. 6.18 (bottom left and bottom right). Additionally to diffusive / dispersive processes, high concentrations of sulfate and low concentrations of methane are transported deeper into the column utilizing pathways which allow "short-cuts", e. g. burrows and small canals formed by animals living in the sediment. This leads to the effect, that methane re-oxidation by sulfate already occurs in lower regions of the sediment column. This time, the re-oxidation area is increased to the upper about 40 % of the domain. In both situations,

6 Applications

sulfide is produced as the resulting product of the re-oxidation process. Note, that for both simulations the concentrations for methane and sulfate at the top boundary are fixed to zero (methane) and the maximum value (sulfate), respectively.

This simple case study underlines the importance to account for possible additional biogeochemical effects such as bioirrigation when dealing with reaction processes in natural environments. Although these effects can only be estimated as they are hard to quantify, they should be taken into account for the simulation of the combined processes of flow, transport and reaction kinetics in subsurface sediments where applicable.

For the modeling of methane concentration distributions at *vent* and *partial vent* sites in Eckernförde Bay, which is presented in the following, the above findings are employed to improve the simulation results.

6.2.3 Model Setup Eckernförde Bay

The following application example confirms that *DiaTrans* is capable of modeling the full set of physical and biogeochemical reaction processes in highly two-dimensional flow which is influenced by density-differences. The setup for the modeling of flow, transport and reaction processes at *vent* and *partial vent* locations in Eckernförde Bay in 2D is displayed in fig. 6.19. The two-dimensional domain represents the uppermost 1 m of the subsurface sediments and is 4 m in width to account for the recirculating fluid flow due to density-differences resulting from inflowing freshwater at the lower boundary. A value of 1 m is chosen again for the z-direction.

The $DIRICHLET$ BCs at the upper boundary are identical to the ones chosen for the case study for the effects of bioirrigation in advective flow in sec. 6.2.2, i.e. a pressure of $p = 300000\,Pa$ for the fluid phase representing a 20 m seawater column and fixed molar concentrations for all dissolved components are used ($C_{mol}^{O_2} = 0.285\,mmol/L$, $C_{mol}^{Cl} = 363\,mmol/L$, $C_{mol}^{NO_3} = 0.01\,mmol/L$, $C_{mol}^{SO_4} = 30.5\,mmol/L$ and $C_{mol}^{Mn^{2+}} = C_{mol}^{Fe^{2+}} = C_{mol}^{H_2S} = C_{mol}^{CH_4} = 0.0\,mmol/L$). At the left and right boundaries, no flow $NEUMANN$ BCs are set for all components.

Submarine groundwater inflow is assumed to occur in the middle part (0.5 m) of the lower boundary, where different freshwater inflow rates q^{H_2O} are assumed, representing a vent location (simulation C) and a partial vent location (simulation D). These values are chosen according to the ones used by HINKELMANN et al. (2002) which have been

6.2 Modeling Reaction Processes at SGD Sites

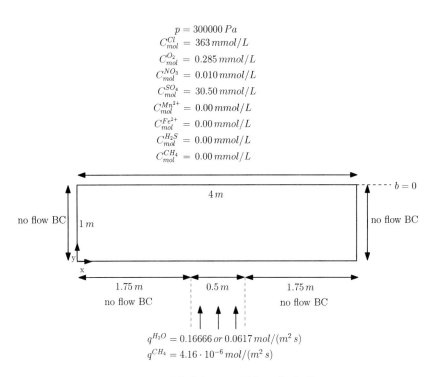

Figure 6.19: Model setup Eckernförde Bay

measured in the field (see table 6.3). The left and right part of this boundary are also set to no flow conditions for all components.

For the boundary conditions representing methane formation at the lower boundary, two major aspects have to be considered. First, the main focus of this application lies on the vertical distribution of methane at the *vent* and *partial vent* locations and second, HINKELMANN et al. (2002) reported that in areas which are not directly influenced by SGD, the maximum solubility of methane is exceeded and gas bubbles form up, evoking an additional gas phase besides the liquid phase. As *DiaTrans* was not designed for simulating multi-phase flow and to avoid unrealistic high solute concentration of methane which would strongly influence the simulation results, the additional high methane formation is only allowed to occur in the same inflow area (0.5 m) where SGD takes place. Therefore, a value of $q^{CH_4} = 4.16 \cdot 10^{-6}\, mol/(m^2\, s)$, which has been reported by HINKELMANN et al. (2002), is set as *NEUMANN* BC for the methane

6 Applications

component at this part of the boundary, whereas no flow condition is assumed as for the other components at the left and right part of the lower boundary.

The assumed exponentially decreasing profiles for the solid concentrations OM, MnO_2 and $Fe(OH)_3$, which are constant over time, are again identical to the ones used during verification (sec. 5.3) and are presented in fig. 5.11.

As initial conditions, the whole domain is completely filled with saltwater with $\rho_{mass} = 1017 \, kg/m^3$, corresponding to $C_{mol}^{Cl} = 363 \, mmol/L$. The initial pressure distribution is hydrostatic and all other molar fractions of dissolved components are set to 0. All relevant physical parameters are summarized in table 6.3.

For simulation C at a *vent* location, only the effects of bioturbation (sec. 3.7) are taken into account, neglecting bioirrigation (sec. 3.8). For simulation D, also a large influence of bioirrigation is regarded with the values described in table 6.3. Again, the transport is governed by density-influenced advective and diffusive / dispersive processes.

As for the case study of bioirrigation in advective flow, methane re-oxidation by O_2 and SO_4 occurs in the uppermost part of the sediment and is modeled using the kinetics for secondary reactions as described in sec. 3.9.1.

Table 6.3: Physical parameters for the simulation at a *vent* (simulation C) and at a *partial vent* (simulation D)

Parameter	simulation C	simulation D
permeability k	$1 \cdot 10^{-11} \, m^2$	
porosity ϕ	0.9	
molecular diffusion coefficient D_m	$1.0 \cdot 10^{-9} \, m^2/s$	
longitudinal dispersion length α_l	$0.01 \, m$	
transversal dispersion length α_t	$0.001 \, m$	
bioirrigation parameters $\alpha_{bi1}/\alpha_{bi2}$	0.0 / 0.0	$10^{-5} \frac{1}{s} / 10 \frac{1}{m}$
submarine groundwater inflow rate q^{H_2O}	$0.16666 \, mol/(m^2 \, s)$	$0.0617 \, mol/(m^2 \, s)$
methane inflow rate q^{CH_4}	$4.16 \cdot 10^{-6} \, mol/(m^2 \, s)$	

A mesh resolution of 25 cells in vertical y-direction and 25 cells in x-direction is applied. The maximum time-step size is 10 minutes and steady-state is reached after approximately 20 days.

6.2 Modeling Reaction Processes at SGD Sites

6.2.4 Results

A typical pressure distribution for simulation C at a *vent* location, which is nearly hydrostatic, is depicted in the upper part of fig. 6.20. Due to the density differences in the system caused by chloride concentration gradients (see 6.20, bottom), a recirculation of the water is induced. Denser saltwater enters the domain across the left and right part of the upper boundary, whereas less dense submarine groundwater, represented by low chloride concentrations, enters the system at the lower boundary and leaves across the middle part of the upper boundary (see lower part in fig. 6.20). The seepage velocity vectors in fig. 6.20, which are not scaled by their magnitude, represent these flow processes. Note, that downward directed velocities are about two to three orders of magnitude smaller than the upward directed ones in the center of the *vent*. Similar results are obtained for simulation D at a *partial vent* location.

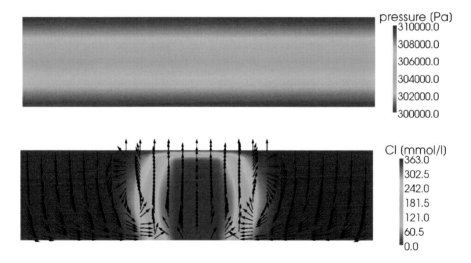

Figure 6.20: Top: Pressure distribution; Bottom: Chloride concentration distribution and seepage velocity vectors (both for simulation C at a *vent* location)

Vent Location

Resulting methane and sulfate concentration distributions for simulation C at a *vent* location can be seen in fig. 6.21. Note, that for this simulation the influence of bioirrigation is neglected, which leads to the fact, that in the center of the *vent* sulfate is only

6 Applications

transported to lower regions by diffusive / dispersive processes. Therefore, the rate of re-oxidation of methane by sulfate is low in this area and methane can reach the upper boundary and leave the system. Zero concentration of methane in the uppermost few centimeters of the domain is only caused by the fact, that methane and sulfate concentrations are fixed at the upper boundary to zero (methane) and to the maximum value (sulfate). Overall, these results confirm the findings from simulation A in sec. 6.2.2.

Only a comparatively low maximum concentration of methane of $1.4\,mmol/L$ occurs. Methane, which enters the system at the lower boundary due to the assumed high formation rate, is transported relatively fast through the system due to the high submarine groundwater discharge advection rate. At *vent* locations, methane concentrations have been measured in the overlying water column close to the sediment-water interface (see SCHLÜTER et al., 2004). This is confirmed by the simulation results, as the maximum methane concentration will reach the upper boundary and enter the seawater column in reality, although methane concentrations are zero at the top boundary in the simulation due to the assumed *DIRICHLET* BC.

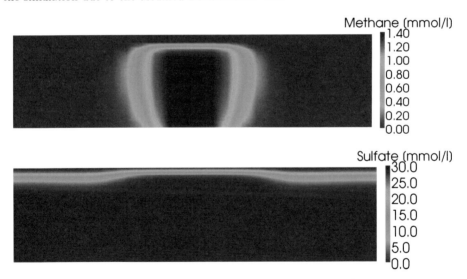

Figure 6.21: Top: Methane concentration distribution; Bottom: Sulfate concentration distribution (both for simulation C at a *vent* location)

In areas of the domain which are almost not influenced by SGD, sulfate is only transported to lower regions by diffusive / dispersive processes and the very small downward

6.2 Modeling Reaction Processes at SGD Sites

directed advective flow which is due to the density differences. Therefore, high sulfate concentrations are only present in the uppermost part throughout the domain.

Partial Vent Location

Fig. 6.22 depicts resulting methane and sulfate concentration distributions for simulation D at a *partial vent* location. In contrast to simulation C, the influence of bioirrigation is regarded here, as it is assumed that this process is a predominant factor at locations less influenced by SGD. This time, sulfate is transported to lower regions in the center of the *partial vent* by bioirrigation additional to diffusive / dispersive processes. Therefore, the rates of re-oxidation of methane by sulfate are a lot higher in this area and methane is completely re-oxidized before it can can reach the upper boundary and leave the system. In nature, it has been measured, that methane concentrations are zero in the surface water column at some locations less influenced by SGD (see SCHLÜTER et al., 2004), which is again confirmed by the simulation results.

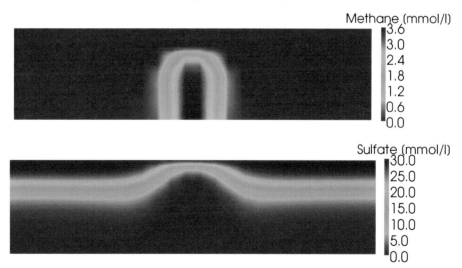

Figure 6.22: Top: Methane concentration distribution; Bottom: Sulfate concentration distribution (both for simulation D at a *partial vent* location)

As seepage velocities and advection rates due to SGD are smaller at *partial vent* locations, methane accumulates to a greater extent in the lower part of the domain and generally higher concentrations occur in the system. For this simulation, a maximum value of $3.6\,mmol/L$ is obtained.

6 Applications

For simulation D, sulfate is generally transported to lower regions by bioirrigation processes additionally to the diffusive / dispersive processes and the very small downward directed advective flow which is due to the density differences. Therefore, higher sulfate concentrations are also present in areas which are not directly influenced by SGD, i. e. in the left and right part of the domain.

Simulation vs. Nature

In the following, the simulation results are compared to field measurements for methane concentration profiles obtained in the course of the *Sub-GATE* project (see SAUTER and SCHLÜTER, 1999). From simulation C at the *vent* location and simulation D at the *partial vent* location, vertical methane concentration profiles are extracted at $x = 2.00\,m$ in the center of the domain, which are displayed in fig. 6.23. The corresponding measurements are also plotted in fig. 6.23.

The simulation results at the *vent* location (C) are in very good agreement with the measurements. The maximum methane concentration of $1.4\,mmol/L$ is nearly identical for simulation and measurements, as well as the shape of the profile, although the rate of methane formation at the lower boundary has never been measured and is only an estimated value. The only major differences occur in the uppermost part of the sediment. This is again due to the fact, that the *DIRICHLET* BCs in the model fix the methane concentration to zero and the sulfate concentration to its maximum value.

Additionally, this good agreement supports the assumption that bioirrigation does not play a predominant role at locations with high influence of submarine groundwater discharge. This might be due to the fact, that irrigating animals avoid areas with relatively high advection rates.

In contrast to simulation C, the results of simulation D differ to a comparatively high extent from the measurements at the *partial vent* location. Especially in lower regions, the model overestimates methane concentrations with the estimated methane formation rate. In regions closer to the sediment-water interface, the results coincide well as both simulation and measurements yield zero concentration of methane (to about $40\,cm$ of depth for the measurements and about $20\,cm$ for the simulation). This again supports the findings from the field, that methane concentrations are negligible in the seawater column. The general shape of the profile is similar, reporting that larger accumulation and therefore higher concentrations of methane occur at greater depths compared to

6.2 Modeling Reaction Processes at SGD Sites

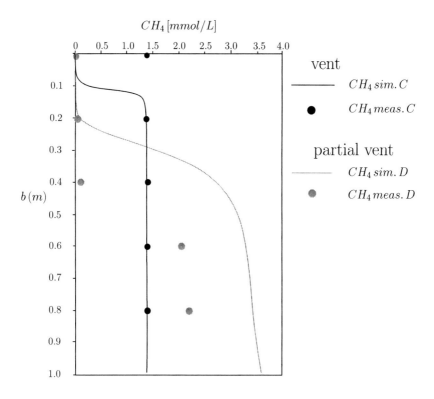

Figure 6.23: Comparison of simulation results (C and D) and measurements for methane and sulfate concentration profiles

the *vent* location. This is due to lower advection rates from submarine groundwater discharge.

SCHLÜTER et al. (2004) presented a summary of methane concentration profiles obtained at various locations in Eckernförde Bay. These profiles, which are displayed in fig. 6.24, have been obtained in areas which are highly influenced by SGD (fig. 6.24, left) and locations less influenced by advective flow (fig. 6.24, right).

When comparing these profiles with the simulation results, an overall agreement in the shape of the profiles can clearly be identified. Most of the profiles at locations highly influenced by SGD are more or less constant over depth, which confirms the modeling results from simulation C. Increasing profiles over depth with very low concentrations of methane in the upper region of the sediment are reported by SCHLÜTER et al.

6 Applications

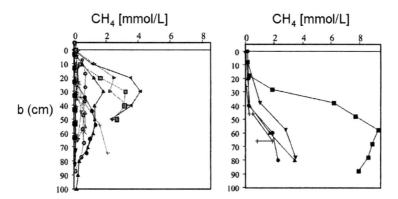

Figure 6.24: Typical methane concentration profiles in Eckernförde Bay at locations highly influenced by SGD (left) and at locations less influenced by advective flow (right) (after SCHLÜTER et al., 2004)

(2004) for less influenced regions and have also been obtained from simulation D. Note, that even higher concentrations of methane than the simulated $3.6\,mmol/L$ have been measured in the field at greater depths, suggesting that the overall trend of the simulation results, at both the *vent* and *partial vent* location, agrees with field measurements.

Generally, a lot of factors, besides uncertainties in field measurements, can influence the simulation results as almost all model parameters are afflicted with uncertainties. In the following, some of these parameters which are believed to have a major effect on the simulation results are summarized:

- The estimated methane formation rate at the lower boundary is believed to have the largest impact on the modeling results. Especially the occurring maximum concentration at the bottom of the domain is highly influenced by this parameter. Actual rates for this methanogenesis at great depths, which are hard to obtain, would certainly improve the results.

- Factors for bioirrigation, which effect simulated concentration profiles especially at *partial vent* sites, are also only estimated.

- Measured profiles for solid concentrations of OC, MnO_2 and $Fe(OH)_3$, instead of the assumed ones, would highly improve the simulation results.

- The values for maximum concentration of sulfate, oxygen, nitrate in the bottom water, which influence the reaction kinetics, may differ to some extent from reality.
- In reality, maximum solubility of methane is exceeded in Eckernförde Bay, as the formation of gas bubbles has been reported. These processes might also have a larger impact on the results than expected.
- Soil parameters such as permeability and porosity are only estimated and assumed to be constant in the whole domain. In reality, the subsurface sediment is highly heterogeneous.
- It is obvious, that dispersion lengths are also only based on values obtained from literature.
- Also, in this study the reaction rates for the kinetics of primary and secondary reactions are only based on available values from literature.

To investigate the influence of the estimated methane formation rate at a *partial vent* site, another simulation is carried out (simulation E), as this parameter is believed to have a major influence on the simulation results. The setup is identical to the one of simulation D, only changing q^{CH_4} to $2.5 \cdot 10^{-6}\, mol/(m^2\, s)$, assuming a lower rate of methanogenesis. A comparison of the methane concentration results of simulations D and E together with the according measurements are displayed in fig. 6.25.

From fig. 6.25 it can clearly be seen, that the influence of the estimated methane formation rate on the simulation results is comparatively large. Although the general shape of the methane concentration profile is similar between simulations D and E, major differences in maximum concentration and extent of the re-oxidation zone can be depicted. The maximum methane concentration in simulation E is now lower at $2.24\, mmol/L$, which agrees well with the according measurements. The extent of the re-oxidation zone is enlarged to the upper $30\, cm$ of the sediment column, which also leads to a better agreement between simulation results and measurements. The comparative results of simulations D and E clearly depict, that the uncertainties of different model parameters have to be carefully investigated in order to obtain satisfactory model results.

The application example for reaction processes at SGD sites in Eckernförde Bay clearly shows, that *DiaTrans* is capable of modeling the full extent of physical and biogeochemical processes even in highly two-dimensional density-driven flow. Overall, the

6 Applications

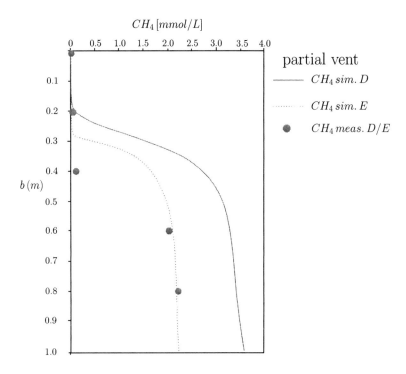

Figure 6.25: Comparison of simulation results (D and E) and measurements for methane concentration profiles

simulation results are reliable and good agreement with field measurements has been obtained.

To improve the simulation results in the future, all of the above mentioned uncertainties for the model parameters could be minimized by providing additional realistic values obtained from extensive field measurements where applicable.

6.3 Modeling Interactions with the Water Column

Numerical modeling can contribute to determine the fate of contaminants or nutrients when crossing the sediment-water interface. The understanding of the controlling physical and biogeochemical processes is of great importance not only in near-shore sediments for different reasons.

On the one hand, dissolved contaminants or nutrients which enter the water column certainly have an influence on the ecology in the near surroundings and therefore might influence benthic habitat conditions. On the other hand, methane, for example, is a greenhouse gas and hence, its pathways in the sediment as well as in the water column is of great interest. Dissolved methane, which is being transported through the sediment column at *vent* locations like in Eckernförde Bay, Baltic Sea, Germany, will eventually enter the overlying water column. If advection rates at a *vent* location are large and the maximum solubility of methane is exceeded in the water column, it may be released as gaseous methane to the atmosphere.

6.3.1 Introduction

In the following, the approach, which allows to give rough qualitative estimates of the fate of constituents, e.g. methane, when entering the water column, is presented. The inclusion of reaction kinetics such as re-oxidation by oxygen or sulfate is possible. The following application example highlights the main interests in this field of research and presents the motivation for further model development.

According to SCHLÜTER et al. (2004), dissolved methane concentrations have been measured in the water column in Eckernförde Bay, Baltic Sea, Germany, which "suggest the release of CH_4 associated with seepage of fluids from the seafloor." The investigation of the fate of dissolved methane in the sediment as well as in the water column is of big interest in this field of research. Additional to high dissolved concentrations, free gas has been reported to be present in the sediment as well as "ebullition of gas bubbles from the seafloor" (SCHLÜTER et al., 2004) which underlines the assumption that methane is released to the atmosphere.

The approach presented herein is employed to estimate the fate of dissolved methane in the bottom water overlying a sediment column at a *vent* location (for details on

6 Applications

vents see sec. 6.2). The idea behind this approach, which can also account for reaction processes in the bottom water, as well as the corresponding model setup are explained in detail in the following. Note, that the approach can be utilized for any dissolved constituents, e. g. contaminants, nutrients or conservative tracers, as it can be adjusted to certain needs.

As with the presented approach only rough qualitative estimates are possible, reasons to promote further development of *DiaTrans*, which is vital to enhance the prediction of the interaction processes at the sediment-water interface and to overcome the deficits of the current approach, are outlined.

6.3.2 Approach & Model Setup

General Setup

The model domain is similar to the one used in sec. 6.2.3 to model reaction processes at *vent* and *partial vent* locations. The model domain is 5 m wide in x-direction, extended to 1.5 m depth in y-direction and 1 m in z-direction. The lower 1.0 m represent the sediment column, whereas the upper 0.5 m stand for the overlying seawater column (see fig. 6.26).

The sediment column can be described as follows. A *vent* location, where submarine groundwater inflow can occur, is located at the lower boundary of the domain at $1.0\,m \leq x \leq 1.5\,m$. The western, eastern and the rest of the bottom boundary are set to no flow *NEUMANN* BCs. At the *vent* location, the permeability of the sediment is higher than in the surroundings, i. e. $k = 1.0 \cdot 10^{-9} m^2$ compared to $k = 1.0 \cdot 10^{-14} m^2$. With this setup it is guaranteed, that all favoured flow paths at the *vent* location are located in the higher permeable area. At the top 0.20 m of the sediment column $(0.8\,m \leq y \leq 1.0\,m)$ even less permeable areas $(k = 1.0 \cdot 10^{-16} m^2)$ are introduced at the locations beside the *vent* $(0 \leq x \leq 1.0\,m$ and $1.5 \leq x \leq 5.0\,m)$, representing low permeable consolidated material. The porosity in the whole sediment column is equal to $\phi = 0.4$.

As *DiaTrans* is generally developed to simulate flow and transport processes in porous media, a very high permeability for the area representing the water column is chosen $(k = 1.0 \cdot 10^{-7} m^2)$. The porosity is set equal to $\phi = 1.0$. The western and eastern

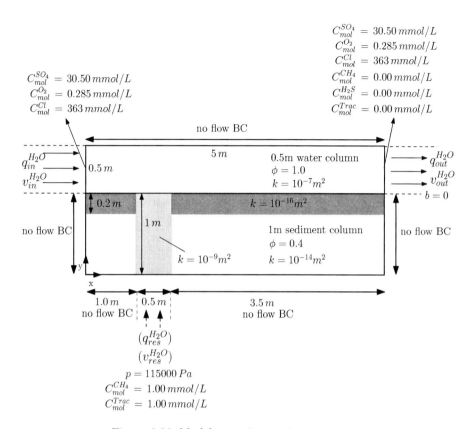

Figure 6.26: Model setup interaction processes

boundaries of the water column are open for inflow and outflow processes, whereas the top boundary is closed utilizing a no flow *NEUMANN* BC.

Inducing a Flow Field in the Water Column

Generally, we are interested in the overall fate of methane in both the sediment and the water column, which is controlled by diffusive/dispersive processes on the one hand and by advective flow and transport processes on the other hand. The advective flux of a component can generally be described as the product of a flow velocity $\underline{v}$ of the fluid phase and the concentration C_{mol}^c. Different possibilities to establish an advective flux in the water column exist. Often, a constant flow field is applied to the western and eastern boundary of the water column to induce a flow, then the advective fluxes are calculated utilizing this flow field.

In the rather simplified approach presented herein, the general mass continuity equation is employed to induce the flow field in the water column. An inflow discharge rate for the liquid phase ($|q_{in}^{H_2O}|$) is applied to the western boundary of the water column as a *NEUMANN* BC. This eventually yields a constant inflow velocity of the water phase ($v_{in}^{H_2O}$) because of continuity reasons, where a flow velocity $\underline{v}$ is equal to a discharge Q divided by the cross-sectional area A. For simplicity reasons, the flow velocities do not vary over depth at the inflow and outflow boundary, respectively.

Note that the resulting inflow velocity is calculated internally in *DiaTrans* utilizing the *Darcy* law, which is employed in the whole domain (i.e. sediment and water column). Although *Darcy*'s law is not valid for flow in free surface water, it is utilized in this approach to produce a constant flow field at the inflow boundary of the water column. Certainly, the resulting pressure distribution in the domain is not the exact physically realistic one, but as mentioned before, we are only interested in the resulting flow velocities to calculate the advective fluxes which govern the transport of a component.

To induce a discharge at the *vent* location entering the sediment, again the principle of mass continuity is employed. At the eastern boundary of the water column, an outflow discharge rate for the water phase is applied as *NEUMANN* BC ($|q_{out}^{H_2O}|$). The difference of inflow and outflow discharge rates ($|q_{out}^{H_2O}|$-$|q_{in}^{H_2O}|$) yields a resulting discharge rate $|q_{res}^{H_2O}|$ which does not have to be explicitly set as *NEUMANN* BC because of continuity reasons. As a constant pressure *DIRICHLET* BC is only used at the bottom of the sediment column at the *vent* location, this resulting discharge stands for the inflow discharge rate at the *vent*.

As for the inflow discharge at the western boundary of the water column, a resulting (*Darcy*) flow velocity is internally calculated at the eastern outflow boundary of the water column ($v_{out}^{H_2O}$) and at the *vent* location ($v_{res}^{H_2O}$) because of the mass continuity reasons described above.

The two parameters $|q_{in}^{H_2O}|$ and $|q_{out}^{H_2O}|$ can now be varied to simulate different scenarios with changing flow velocities at the inflow boundary of the water column and at the *vent* location.

It is obvious, that this approach can only be used to roughly estimate the fate of constituents in the water column. It is only employed for the following simulations

6.3 Modeling Interactions with the Water Column

to illustrate the main interests in this field of research. For a fully physically realistic representation of the occurring interaction processes, further development of *DiaTrans*, e.g. by introducing the shallow-water equations, is certainly necessary.

Components and Reaction Processes

As the calculations are carried out with density-dependent flow and transport processes, chloride (Cl) is considered as a component beside pure fresh water (H_2O). A maximum saltwater density of $\rho_{mass} = 1017\,kg/m^3$ is considered, corresponding to a maximum chloride concentration of $C_{mol}^{Cl} = 363\,mmol/L$. Initially, the whole domain is filled with saltwater. As the flow induced in the water column is assumed to consist of pure saltwater, maximum chloride concentrations are also set at the inflow and outflow boundaries of the water column as *DIRICHLET* BCs.

As secondary reaction processes of methane (CH_4), which are allowed to occur in the whole domain, are considered, also sulfate (SO_4), oxygen (O_2) and sulfide (H_2S) are included for the simulations as those components influence the reaction kinetics of methane directly. Methane is assumed to be absent in the whole domain initially. A maximum concentration of methane of $C_{mol}^{CH_4} = 1.0\,mmol/L$ is set at the bottom of the domain only at the *vent* location (*DIRICHLET* BC). This value is similar to results of sec. 6.2, where it has been observed that a maximum concentration of $1.4\,mmol/L$ will reach the sediment-water boundary. At the eastern outflow boundary, the methane concentration is set to $0.0\,mmol/L$ (*DIRICHLET* BC).

On the contrary, the part of the domain which represents the water column, is initially filled with the maximum concentration of sulfate ($C_{mol}^{SO_4} = 30.5\,mmol/L$) as well as oxygen ($C_{mol}^{O_2} = 0.285\,mmol/L$), representing saltwater in the Baltic Sea. As for chloride, these maximum concentrations are also set as *DIRICHLET* BCs at the inflow and outflow area of the water column. The concentration for sulfide is set to $0.0\,mmol/L$ initially, as well as at the outflow boundary.

As the fate of methane in the water column is of interest in this application, selected secondary reactions are considered, neglecting the primary reactions occurring in the sediment. The chosen reactions to account for the main re-oxidation processes of methane including by-products are:

- Sulfide Re-Oxidation by O_2 (eq. 3.33)

6 Applications

- Methane Re-Oxidation by O_2 (eq. 3.34)
- Methane Re-Oxidation by SO_4^{2-} (eq. 3.35)

Note, that resulting reaction rates for the secondary reactions are higher in the water column, as these reactions occur in the available pore space, which is described by the porosity ϕ. This value is assumed to be 1.0 in the water column, compared to 0.4 in the sediment.

For comparison, a conservative tracer ($M^{Trac} = 0.018\,kg/mol$) is also included as a component to investigate the effect of the reaction processes on methane concentration distributions. A maximum concentration of $C_{mol}^{Trac} = 1.0\,mmol/L$ is assumed at the bottom of the domain only at the *vent* location similar to methane (*DIRICHLET* BC). Initially, the tracer is absent in the whole domain. At the eastern outflow boundary, the tracer concentration is set to $0.0\,mmol/L$ (*DIRICHLET* BC).

Scenarios

Three different scenarios are considered, namely A, B and C (see table 6.4). In these scenarios, all parameters are identical, except the inflow rate $q_{in}^{H_2O}$, the outflow rate $q_{out}^{H_2O}$ and the resulting *vent* flow rate $q_{res}^{H_2O}$. The flow rates are chosen in such a way, that the flow velocities at the inflow area of the water column and at the *vent* location show distinct ratios.

For scenario A, the resulting inflow velocity in the water column is about twice as big as the one in the *vent*; for scenario B, the flow velocities are identical and for scenario C, the inflow velocity in the water column is about half as big as the one in the *vent* (see table 6.4). As HINKELMANN et al. (2002) stated, that flow velocities inside a *vent* can be as big as $1\,cm/s$ and therefore be in the range of surface water velocities, the flow velocity in the *vent* is chosen to be $0.5\,cm/s$, corresponding to $q_{res}^{H_2O} = 300\,mol/(m^2\,s)$. The flow velocities in the inflow area of the water column are $0.0025\,m/s$, $0.0050\,m/s$ and $0.010\,m/s$, respectively. Note that the seepage velocities ($v_{res,s}^{H_2O}$) in the *vent*, which govern advective transport, result from dividing the *Darcy* flow velocity ($v_{res}^{H_2O}$) depicted in table 6.4 by the sediment porosity ($\phi = 0.4$).

The molecular diffusion coefficient D_m is assumed to be identical for all components and dispersion lengths as shown in table 6.4 are employed. Note, that dispersive fluxes in addition to diffusive fluxes are also calculated in the water column, although dispersion

6.3 Modeling Interactions with the Water Column

Table 6.4: Parameters for scenarios A, B and C for the simulation of interaction processes

Parameter	scenario A	scenario B	scenario C
inflow rate $|q_{in}^{H_2O}|$	$600\,mol/(m^2\,s)$	$300\,mol/(m^2\,s)$	$150\,mol/(m^2\,s)$
inflow velocity $v_{in}^{H_2O}$	$\sim 0.010\,m/s$	$\sim 0.005\,m/s$	$\sim 0.0025\,m/s$
outflow rate $|q_{out}^{H_2O}|$	$900\,mol/(m^2\,s)$	$600\,mol/(m^2\,s)$	$450\,mol/(m^2\,s)$
outflow velocity $v_{out}^{H_2O}$	$\sim 0.015\,m/s$	$\sim 0.010\,m/s$	$\sim 0.0075\,m/s$
vent flow rate $|q_{res}^{H_2O}|$	$300\,mol/(m^2\,s)$	$300\,mol/(m^2\,s)$	$300\,mol/(m^2\,s)$
vent Darcy velocity $v_{res}^{H_2O}$	$\sim 0.005\,m/s$	$\sim 0.005\,m/s$	$\sim 0.005\,m/s$
vent seepage velocity $v_{res.s}^{H_2O}$	$\sim 0.0125\,m/s$	$\sim 0.0125\,m/s$	$\sim 0.0125\,m/s$
max. methane conc. $C_{mol}^{CH_4}$		$1.0\,mmol/L$	
max. tracer conc. C_{mol}^{Trac}		$1.0\,mmol/L$	
mol. diffusion coefficient D_m		$1.0 \cdot 10^{-9}\,m^2/s$	
longit. dispersion length α_l		$0.01\,m$	
transv. dispersion length α_t		$0.001\,m$	
bioirrig. parameters $\alpha_{bi1}/\alpha_{bi2}$		$10^{-5}\,\frac{1}{s}\,/\,10\,\frac{1}{m}$	

is a process solely occurring in porous media (see sec. 3.5.2). The additional admixture induced by dispersion in the water column is assumed to qualitatively substitute turbulence effects in the water column, which also enhance mixing.

All scenarios are carried out with bioturbation (see eq. 3.23 for parameters) and bioirrigation effects enabled (see table 6.4 for parameters). *DiaTrans* automatically accounts for the fact, that these processes are only occurring in the subsurface.

The model domain is discretized with 40 elements in x-direction and 50 elements in y-direction. Mesh convergence is reached. The maximum time-step size is 10 seconds due to the large advection rates and a steady-state situation is obtained after about 20 to 60 minutes, depending on the scenario.

In the following, exemplary simulation results for the concentration distributions of methane and the conservative tracer are presented and discussed in detail.

6.3.3 Results

A typical *Darcy* (flow) velocity distribution is shown in fig. 6.27 (here illustrated for scenario B). Note, that the solid black line indicates the sediment-water interface and that the lengths of the flow velocity vectors are scaled by the velocity magnitude. It can clearly be seen that all favoured flow paths in the sediment are located in the higher

6 Applications

permeable areas at the *vent* location. The velocities in the lower permeable sediment areas are negligible and therefore not visible due to the scaling.

The advective flow entering the water column from the *vent* influences the horizontally aligned flow in the bottom water in the vicinity of the *vent*. As mentioned before, due to continuity reasons of the simplified approach, the flow velocities are higher at the eastern boundary of the water column. Similar flow patterns can be obtained for scenarios A and C.

Figure 6.27: *Darcy* (flow) velocity distribution (Scenario B)

Scenario A

The resulting methane concentration distribution for scenario A is shown in fig. 6.28 (top). Again, the solid black line indicates the sediment-water interface. Special attention should be paid to the flow velocities at the inflow area of the water column compared to the velocities occurring in the *vent*. As the inflow velocity determines how fast maximum sulfate and oxygen concentrations are transported to the *vent* location in the water column, the ratio of this flow velocity to the velocity in the *vent* controls the re-oxidation processes of methane by both sulfate and oxygen.

In scenario A, the maximum inflow velocity is about twice as big as the resulting *Darcy* flow velocity in the *vent* and therefore almost as big as the resulting seepage velocity. Methane, which is transported towards the sediment-water interface in the sediment is almost completely re-oxidized by sulfate and oxygen carried towards the *vent* location from the western inflow boundary in the water column. Therefore, only low concentrations of methane are present in the vicinity of the *vent* in the water column for this scenario.

6.3 Modeling Interactions with the Water Column

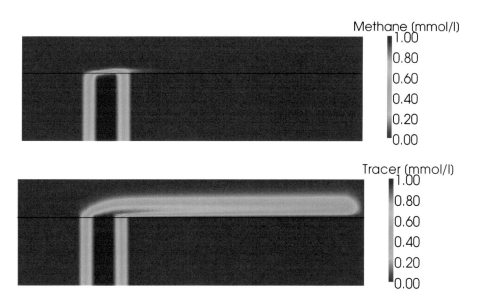

Figure 6.28: Concentration distributions for methane and a conservative tracer (Scenario A)

The influence of the secondary reaction processes can clearly be depicted when comparing the resulting methane concentration distribution (fig. 6.28, top) to the concentration distribution of a conservative tracer, which is not influenced by the reaction kinetics (fig. 6.28, bottom).

It is obvious, that higher concentrations of the tracer occur in the water column. Although the tracer concentrations get diluted and dispersed in the water column, average concentrations of $0.5\,mmol/L$ reach about the middle of the water column in y-direction. Generally, tracer concentrations are swept through the system towards the eastern water column boundary. Note, that the tracer concentrations are fixed to $0.0\,mmol/L$ at the outflow boundary of the water column.

This basic comparison clearly highlights the importance to also regard reaction kinetics when investigating the fate of contaminants or nutrients at the sediment-water interface. When these biogeochemical processes are considered, improved estimates can be obtained.

6 Applications

Scenario B

The influence of a lower inflow velocity into the water column is outlined in fig. 6.29. Again, the solid black line indicates the sediment-water interface and the main focus should lie on the area reaching from the western water column boundary to the vicinity around the *vent*. In this scenario, the *Darcy* flow velocity inside the *vent* is about as big as the velocity in the overlying seawater. Therefore, the seepage velocity in the *vent* is more than twice as big as the inflow velocity (compare to table 6.4). As sulfate and oxygen concentrations, which are transported to the *vent* location from the western water column boundary, do not completely re-oxidize the upcoming methane, average methane concentration of about $0.5\,mmol/L$ reach a height of approximately $10\,cm$ in the water column.

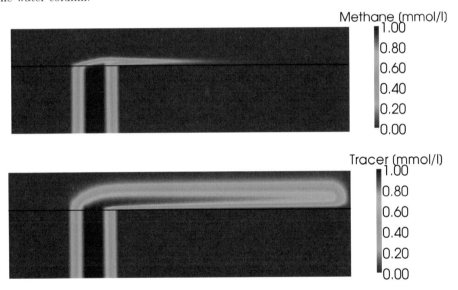

Figure 6.29: Concentration distributions for methane and a conservative tracer (Scenario *B*)

Again, the influence of the reaction processes can be depicted, when comparing resulting methane concentrations (fig. 6.29, top) to the concentration distribution of the conservative tracer, which is not influenced by the reaction kinetics (fig. 6.29, bottom). The average tracer concentration reaches about $75\,\%$ of the height of the water column. Again, the tracer concentrations are fixed to $0.0\,mmol/L$ at the eastern water column boundary due to the *DIRICHLET* BC.

6.3 Modeling Interactions with the Water Column

Scenario C

Scenario C represents a situation, in which the advective flux of upcoming methane clearly dominates the system. Again, the solid black line indicates the sediment-water interface. The *Darcy* flow velocity in the *vent* is now about twice as big as the inflow velocity of the water column, corresponding to a seepage velocity which is about five times bigger. As sulfate and oxygen are transported slower towards the *vent* location in the water column, the rates of re-oxidation processes are smaller. This results in higher concentrations of methane in the water column (see fig. 6.30, top). Again, one can compare to the resulting tracer concentrations (fig. 6.30, bottom) to clearly see the influence of the reaction processes. Very low tracer concentrations arrive at the top boundary, which indicates that in reality tracer concentrations would be present at the water-atmosphere interface. Again, the tracer concentrations are fixed to $0.0\,mmol/L$ at the eastern water column boundary due to the *DIRICHLET* BC.

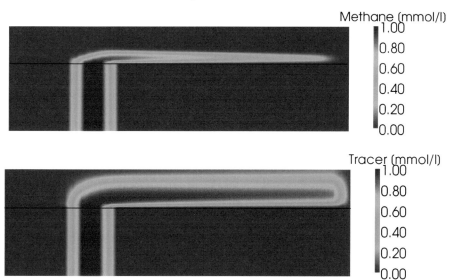

Figure 6.30: Concentration distributions for methane and a conservative tracer (Scenario C)

For this case, the average methane concentration of $0.5\,mmol/L$ is able to reach heights of about $20\,cm$ in the water column and generally higher concentrations are present in the system. One can clearly imagine, that at even higher discharge rates and therefore flow velocities in the *vent* or with higher initial concentrations of methane, methane will

6 *Applications*

eventually reach the water-atmosphere interface where it might be released in gaseous form when the maximum solubility is exceeded. The understanding and estimation of such processes is of great importance in this field of research.

Also lower water depths of the overlying bottom water would intensify this process, as dissolved methane only has to pass a shallower water body. This aspect is especially important, when dealing with tidal effects. During low tide, not only the water depths are shallow but also submarine groundwater discharge rates, which may transport methane, are generally higher due to a larger pressure gradient which automatically results from the smaller water depths.

Generally, numerical modeling can be very useful to improve the understanding of the processes occurring at the sediment-water interface, not only in near-shore sediments. To illustrate the main interests in this field, a simplified approach has been presented, which allows to use *DiaTrans* for qualitative estimates about the fate of contaminants or nutrients when entering the water column. The application example clearly shows that it is vital to include relevant reaction processes also in the field of sediment-water interaction, especially when talking about components such as methane which are highly influenced by other constituents, e. g. sulfate and oxygen in this case.

The major deficit of this approach is its calculation of the flow field in the water column, which is not completely physically realistic. This application should only be understood as an illustrative example and as a "door-opener" for further development of *DiaTrans* to assess the interaction processes at the sediment-water interface.

7
Summary, Conclusions & Outlook

As the investigation of the transition zone between subsurface sediments and the seawater column becomes more and more important when dealing with ecological questions not only in coastal areas, the numerical modeling of flow, transport and reaction processes in the sediment and across the sediment-water interface is one of the main objectives in this field of research. Other objectives are the quantification of the nutrient recycling, of competing bacterial processes affecting the decomposition of organic matter and of the admixture of chemical constituents into the seawater column with special focus on the water quality.

DiaTrans – a multi-component model for density-driven flow, transport and biogeochemical reaction processes in the subsurface has been presented in this work. Details about the model development, which has been the main objective, have been outlined and application examples for physical and biogeochemical processes have been given. These include the modeling of physical processes and biogeochemical reaction processes at submarine groundwater influenced sites.

In the following, the contents of this work are summarized, further conclusions are drawn and an outlook on further possible development is given.

Chapter 1

An introduction, including the overall motivation of the work has been presented in this chapter. The phenomenon of submarine groundwater discharge, which has been observed in the North Sea and Baltic Sea, Germany, for example, has been explained in detail. Also, an overview of the deficits of current numerical models in the field of subsurface flow, transport and reaction modeling has been presented.

Summary, Conclusions & Outlook

Chapter 2

The main study area in the Wadden Sea of Cuxhaven, North Sea, Germany, where the nature of *sand boils* has been investigated in detail, has been depicted. Measurement techniques employed by the *Alfred Wegener Institute for Polar and Marine Research*, Bremerhaven, Germany to obtain porewater concentration profiles or submarine groundwater discharge rates have been explained.

For even better possibilities to verify *DiaTrans* with field measurements, additional measurement campaigns should be carried out, especially to gather more information about physical and biogeochemical parameters. More details about the correlation of saltwater heights in the Wadden Sea and corresponding submarine groundwater discharge rates would certainly improve simulation results of tidal influenced processes. Similarly, more information about the reaction kinetics and rates would enhance the simulations dealing with biogeochemical processes.

Chapter 3

In this chapter, the underlying model concepts of *DiaTrans* have been explained in detail. Besides information about the continuum approach for porous media and about fluid and soil properties, this includes the background of physical (advection and diffusion / dispersion) and biogeochemical processes (bioturbation, bioirrigation, reactions) occurring in subsurface sediments.

Although typical processes only occurring in coastal environments such as bioturbation and bioirrigation are incorporated in *DiaTrans*, the model can generally be applied to a wide field of applications. Therefore, the simulation of flow, transport and reaction processes is not limited to coastal environments as *DiaTrans* can be utilized for subsurface flow, transport and reaction problems in porous media in general, also on larger scales.

A detailed explanation oft the one-phase / multi-component formulation employed in this work is also presented. Current models for the simulation of biogeochemical reaction processes are often of one-dimensional nature and only employ a simplified flow field. As especially at submarine groundwater influenced sites advective processes cannot be neglected, this multi-dimensional fully coupled approach, including density effects on flow and transport processes is a new approach in this field which also allows

Summary, Conclusions & Outlook

to account for local heterogeneities of the sediment column. Therefore, more realistic estimates for the quantification of dissolved component fluxes in subsurface sediments are possible.

Due to its object-oriented modular structure, *DiaTrans* can be extended to additional physical processes (e.g. *Darcy-Brinkmann* equation) or biogeochemical reactions in a simple manner in the future. As the formulation is not restricted to the number of dissolved components, the model can easily be adjusted to fit certain needs, e.g. the amount of primary and secondary biogeochemical reactions can be chosen according to the problem to be solved.

Chapter 4

The numerical methods employed in *DiaTrans* as well as their implementation has been outlined. The main advantages and disadvantages of the programming language *Java* are depicted, followed by details about the *Finite-Volume-Method*. In *DiaTrans*, this method is applied to rectangular structured grids and the possibility to carry out simulations either on horizontal or vertical planes is given. For further development, local refinement of the computational mesh could easily be implemented which would save computational time. Additionally, an even more detailed representation of the occurring processes, especially of the reaction kinetics, would be possible.

The discretization and implementation techniques for all physical and biogeochemical terms are also explained in detail, with special focus on the upwinding-technique for the advection term. For future work, higher order schemes such as slope-limiter or flux-limiter methods could be tested. Although the fully-upwinding technique has been proven to be reliable in this field of applications, the numerical stability and reliability of the model, especially for numerically demanding simulations including a large number of components and biogeochemical reaction kinetics, might be enhanced.

Information about techniques to solve highly nonlinear, non-symmetric and sparse systems of equations is given, with special focus on the *Newton-Raphson* method which is utilized in this work. Due to the object-oriented nature of *DiaTrans*, this method automatically adjusts to the specified problem and is therefore invariant to the number of regarded components for certain applications.

After values for the minimum and maximum time-step size have been specified by the user, the adaptive time-stepping technique presented herein always uses the largest numerically possible time-step to save computational time. For further development, this technique could be enhanced by implementing a physically-biogeochemically based approach. With such a technique the time-stepping could account for the fully coupled physical and biogeochemical processes. For example, the *Damköhler* number, which relates chemical reaction timescales to other phenomena occurring in a system such as advection, could be employed.

For bigger scale problems with a large number of unknowns, numerical parallelization techniques to minimize computational time could be used, which can easily be done for structured meshes.

Chapter 5

Verification examples for flow, transport and reaction processes in the subsurface have been presented. This includes the *Henry* problem, the *salt dome* problem and standard reaction processes in columns. *DiaTrans'* performance in those verification examples has been discussed in detail.

For the *Henry* problem, the results obtained with *DiaTrans* are in very good agreement with *Segol*'s revised solution and the results obtained with other numerical models. *DiaTrans'* results underline and confirm its strong capability to simulate horizontally driven, density-dependent flow and transport very reliable.

Also for the modeling of density-driven flow and transport mainly in the vertical direction, which is numerically more demanding, *DiaTrans* yields good results as shown for the *salt dome* problem. The model is capable of reproducing the results of well accepted numerical codes, such as *FEFLOW* or *ROCKFLOW*, accurately.

The results of both the *Henry* and the *salt dome* problem show, that the numerical code inside *DiaTrans* is well developed and tested for density-driven flow and transport problems in subsurface sediments.

DiaTrans has also been verified for the modeling of primary and secondary reaction processes. Although no typical benchmark tests exist for this kind of problem, it has been presented that the model is highly capable of reproducing the results of other authors qualitatively. These results demonstrate, that the numerical code inside *DiaTrans* is also well developed for biogeochemical reaction processes.

Chapter 6

Different application examples have been presented. This includes the simulation of physical processes at so-called *sand boil* sites, comprising the tidal effect on porewater concentration profiles. Also, biogeochemical reaction processes at submarine groundwater discharge sites are investigated, with special focus on methane concentration distribution at so-called *vent* and *partial vent* sites. In the last application example, a simplified approach to investigate the physical and biogeochemical processes across the sediment-water interface is employed.

The main goal of the extensive study on the simulation of physical processes at *sand boils* has been to gain a better understanding of and an enhanced insight into these processes. A sensitivity analysis of the model and physical parameters is included, as well as a comparison of model results and field measurements of chloride concentration profiles.

The tidal influence on submarine groundwater discharges and chloride concentration profiles has also been investigated. Special attention was paid to maximum freshwater fluxes occurring during tidal cycles.

Although the flow through *sand boils* is more or less radial-symmetric in nature, the results of the above mentioned preceding investigations, which have been carried out using the simulation tool *MUFTE_UG*, have made it evident that a two-dimensional cross-sectional model for density-driven flow and transport in the subsurface is highly capable of reflecting major findings which have been observed in the field.

A simple comparative study of *DiaTrans* and *MUFTE_UG* showed, that *DiaTrans* is capable of representing the flow and transport processes at *sand boils* occurring in nature, although its reliability could further be verified with additional field data in the future.

The application example for reaction processes at SGD sites in Eckernförde Bay have clearly shown, that *DiaTrans* is also capable of modeling the full extent of physical and biogeochemical processes even in highly two-dimensional density-driven flow. The simulation results, which are reliable and in good agreement with field measurements should also be verified in the future. To improve the simulation results, uncertainties of the model parameters, such as estimated methane formation rate, should be minimized by providing additional realistic values obtained from extensive field measurements.

Summary, Conclusions & Outlook

The last application example deals with interactions of the sediment and the overlying water column, as generally, numerical modeling can be very useful to improve the understanding of the processes occurring at the sediment-water interface, not only in near-shore sediments. To illustrate the main interests in this field, a simplified approach has been presented, which allows to use *DiaTrans* for rough qualitative estimates about the fate of contaminants or nutrients when entering the water column. The application example has underlined that it is important to include relevant reaction processes also in the field of sediment-water interaction, as the fate of components such as methane is highly influenced by the presence of other constituents.

The major deficit of the presented approach is its calculation of the flow field in the water column, which is not completely physically realistic. As this application should only be understood as an illustrative example, the main focus for further development of *DiaTrans* should clearly lie on the assessment of these interaction processes. Either *DiaTrans* could be coupled to other free-surface water models or the model could be extended by including the shallow-water equations explicitly.

Bibliography

ALLER, R. C. (1980), 'Diagenetic Processes Near the Sediment-Water Interface of Long Island Sound. I and II', *Advances in Geophysics* **22**.

ANDERSON, E., BAI, Z., BISCHOF, C., BLACKFORD, S., DEMMEL, J., DONGARRA, J., DU CROZ, J., GREENBAUM, A., HAMMARLING, S., MCKENNEY, A. and SORENSEN, D. (1999), *LAPACK Users' Guide*, Third edn, SIAM (Society for Industrial and Applied Mathematics), Philadelphia, U.S.

BARLAG, C., HINKELMANN, R., HELMIG, R. and ZIELKE, W. (1998), Adaptive Methods for Modelling Transport Processes in Fractured Subsurface Systems, *in* '3rd International Conference on Hydroscience and Engineering, Cottbus, Center of Computational Hydroscience and Engineering, The University of Mississippi'.

BASTIAN, P., JOHANNSEN, K., NEUSS, N., RENTZ-REICHERT, H., WAGNER, C. and WITTUM, G. (1995), *ug 3.0, tutorial, revised edition (April 1995)*, Stuttgart, Germany.

BEAR, J. (1972), *Dynamics of Fluids in Porous Media*, Environmental Sciences Series, Elsevier, New York.

BLACKFORD, L. S., DEMMEL, J., DONGARRA, J., DUFF, I., HAMMARLING, S., HENRY, G., HEROUX, M., KAUFMAN, L., LUMSDAINE, A., PETITET, A., POZO, R., REMINGTON, K. and WHALEY, R. C. (2002), 'An updated set of Basic Linear Algebra Subprograms (BLAS)', *ACM Transactions on Mathematical Software* **28**(2), 135–151.

BLANCH, H. W. (1981), 'Invited review: Microbial growth kinetics', *Chemical Engineering Communications* **8**(4-6), 181–211.

Bibliography

BOKUNIEWICZ, H. (1980), 'Groundwater Seepage into Great South Bay, New-York', *Estuarine and Coastal Marine Science* **10**(4), 437–444.

BOUDREAU, B. P. (1996), 'A method-of-lines code for carbon and nutrient diagenesis in aquatic sediments', *Computers & Geosciences* **22**(5), 479–496.

BOUDREAU, B. P. (1997), *Diagenetic Models and their Implementation*, Springer.

BOUDREAU, B. P. and MARINELLI, R. L. (1994), 'A modelling study of discontinuous biological irrigation', *Journal of Marine Research* **52**, 947–968.

BOUDREAU, B. P. and WESTRICH, J. T. (1984), 'The dependence of bacterial sulfate reduction on sulfate concentration in marine sediments', *Geochemica et Cosmochimica Acta* **48**(12), 2503–2516.

BREIER, J. (2006), The impact of groundwater flows on estuaries, *in* 'Aquifers of the Gulf Coast of Texas', 365. Texas Water Development Board, Austin, Texas, pp. 165–172.

BRUSSINO, G. and SONNAD, V. (1989), 'A Comparison of Direct And Preconditioned Iterative Techniques for Sparse, Unsymmetric Systems of Linear-Equations', *International Journal For Numerical Methods In Engineering* **28**(4), 801–815.

BULL, J. M., SMITH, L. A., BALL, C., POTTAGE, L. and FREEMAN, R. (2003), 'Benchmarking Java against C and Fortran for scientific applications', *Concurrency and Computation-Practice & Experience* **15**(3-5), 417–430. Acm 2001 Java Grande/International Symposium On Computing in Object-Oriented Parallel Environments, Stanford, California, Jun 02-04, 2001.

BURNETT, W. C., BOKUNIEWICZ, H., HUETTEL, M., MOORE, W. S. and TANIGUCHI, M. (2003), 'Groundwater and pore water inputs to the coastal zone', *Biogeochemistry* **66**(1-2), 3–33.

BUSSMANN, I., DANDO, P. R., NIVEN, S. J. and SUESS, E. (1999), 'Groundwater seepage in the marine environment: role for mass flux and bacterial activity', *Marien Ecology - Progress Series* **178**, 169–177.

CHARETTE, M. (2004), 'Water Flowing Underground - new techniques reveal the importance of groundwater seeping into the sea', *Oceanus - the online magazine of research from woods hole oceanographic institution* **43**(1), 29–33.

CHOUDHARI, P. (2001), 'Java Advantages & Disadvantages', Webpage: http://arizonacommunity.com/articles/java_32001.shtml. Last Checked 2001.

CHRISTENSEN, J. P., DEVOL, A. H. and SMETHIE, W. M. (1984), 'Biological enhancement of solute exchange between sediments and bottom water on the Washington continental shelf', *Continental Shelf Research* **3**, 9–23.

CIRPKA, O. A. (1997), Numerische Methoden zur Simulation des reaktiven Mehrkomponententransports im Grundwasser, Dissertation, Institut für Wasserbau, Universität Stuttgart, Stuttgart, Germany.

CIRPKA, O. A. (2005), Ausbreitungs- und Transportvorgänge in Strömungen II - Stoff- und Wärmetransport in natürlichen Hydrosystemen. Lecture Notes, University of Stuttgart.

CLASS, H. (2004), 'Applikation von Ein- und Mehrphasenmodellen für umweltrelevante und technische Fragestellungen', Lecture Notes, Institute of Hydraulic Engineering, University Stuttgart, Stuttgart.

DE JOSSELIN DE JONG, G. (1958), 'Longitudinal and transverse diffusion in granular deposits', *Trans. Amer. Geophys. Union* **39**, 67–74.

DE STADELHOFEN, C. M. (1995), *Anwendung geophysikalischer Methoden in der Hydrogeologie*, Springer-Verlag, Berlin, Heidelberg.

DIERSCH, H.-J. (1994), *FEFLOW User's Manual Version 4.2*, WASY GmbH, Berlin, Germany.

DIERSCH, H.-J. (1995), *FEFLOW Reference Manual*, WASY GmbH, Berlin, Germany.

DIERSCH, H.-J. G. and KOLDITZ, O. (2005), Variable-density flow and transport in porous media: approaches and challenges, *in* 'WASY Software, FEFLOW White Papers Vol. II', WASY GmbH Institute for Water Resources Planning and Systems Research, Waltersdorfer Straße 105, 12526 Berlin, Germany, pp. 5–102.

DIETRICH, E. (2006), Entwicklung eines modularen Systems zur Messung von Ausstromraten in Küsten und Tiefseeregionen, Bachelor thesis, University of Applied Science, Bremerhaven.

EMERSON, S. and HEDGES, J. (2003), 'Sediment Diagenesis and Benthic Flux', *Treatise on Geochemistry* **6**, 293–319.

EMERSON, S., JAHNKE, R., BENDER, M., FROELICH, P., KLINKHAMMER, G., BOWSER, C. and SETLOCK, G. (1980), 'Early Diagenesis In Sediments from the Eastern Equatorial Pacific .1. Pore Water Nutrient and Carbonate Results', *Earth and Planetary Science Letters* **49**(1), 57–80.

EMERSON, S., JAHNKE, R. and HEGGIE, D. (1984), 'Sediment-Water Exchange In Shallow-Water Estuarine Sediments', *Journal Of Marine Research* **42**(3), 709–730.

FAIRES, D. and BURDEN, R. L. (1998), *Numerical Methods*, Thomson Learning.

FESEKER, T. (2004), Numerical Studies on Groundwater Flow in Coastal Aquifers, Dissertation, Universität Bremen, Bremen.

FOSSING, H., BERG, P., THAMDRUP, B., RYSGAARD, S., SØRENSEN, H. and NIELSEN, K. (2004), A model set-up for an oxygen and nutrient flux model for Aarhus Bay (Denmark), NERI Technical Report No. 483, National Environmental Research Institute, Denmark.

FREEZE, R. A. and CHERRY, J. A. (1979), *Groundwater*, Prentice Hall, Englewood Cliffs, New Jersey.

FRIND, E. and MOLSON, J. (1994), *SALTFLOW 2.0 User Guide*, Waterloo Centre for Groundwater Research, Waterloo, Canada.

FROELICH, P. N., KLINKHAMMER, G. P., BENDER, M. L., LUEDTKE, N. A., HEATH, G. R., CULLEN, D., DAUPHIN, P., HAMMOND, D., HARTMAN, B. and MAYNARD, V. (1979), 'Early oxidation of organic matter in pleagic sediments of the eastern equatorial Atlantic: Suboxic diagenesis', *Geochim. et Cosmochim. Acta* **44**, 1075–1090.

HAESE, R. R. (1999), The Biogeochemistry of Iron, *in* H. D. SCHULZ and M. ZABEL, eds, 'Marine Geochemistry', Springer-Verlag, Berlin, Heidelberg, pp. 241–270.

HARBAUGH, A. W. (2005), MODFLOW-2005, The U.S. Geological Survey Modular Ground-Water Model - the Ground-Water Flow Process, U.S. Geological Survey Techniques and Methods 6-A16.

HEIMSUND, B.-O. (2008), 'Matrix Toolkit for Java (MTJ)', Webpage: http://ressim.berlios.de/. Last Checked November, 2008.

HELMIG, R. (1997), *Multiphase flow and transport processes in the subsurface: A contribution to the modeling of hydrosystems*, Springer, Berlin, Heidelberg.

HELMIG, R., CLASS, H., HUBER, R., SHETA, H., EWING, J., HINKELMANN, R., JAKOBS, H. and BASTIAN, P. (1998), 'Architecture of the Modular Program System MUFTE-UG for Simulating Multiphase Flow and Transport Processes in Heterogeneous Porous Media', *Mathematische Geologie* **2**, 123–131.

HENRY, H. R. (1964), Effects of dispersion on salt encroachment in coastal aquifers, *in* H. H. Cooper et al., ed., 'Sea Water in Coastal Aquifers', US Geological Survey Water Supply Paper 1613-C.

HERBERT, A. W., JACKSON, C. P. and LEVER, D. A. (1988), 'Coupled groundwater-flow and solute transport with fluid density strongly dependent upon concentration', *Water Resources Research* **24**(10), 1781–1795.

HINKELMANN, R. (2005), *Efficient Numerical Methods and Information-Processing Techniques for Modeling Hydro- and Environmental Systems*, Vol. 21 of *Lecture Notes in Applied and Computational Mechanics*, Springer, Berlin, Heidelberg, New York.

HINKELMANN, R. and HELMIG, R. (2002), Numerical Modelling of Transport Processes in the Subsurface, *in* G. Wefer, D. Billett, D. Hebbeln, B. Jorgensen, M. Schlüter and T. van Wering, eds, 'Ocean Margin Systems', Springer-Verlag, Berlin, Heidelberg, pp. 269–294.

HINKELMANN, R., SHETA, H., CLASS, H. and HELMIG, R. (2000a), A Comparison of Different Model Concepts for Saltwater Intrusion Processes, *in* F. Stauffer, W. Kinzelbach, K. Kovar and E. Hoehn, eds, 'ModelCARE99: Calibration and Reliability in Groundwater Modelling - Coping with Uncertainties', Vol. 265, IAHS Publications, pp. 385–391.

HINKELMANN, R., SHETA, H., CLASS, H., SAUTER, E. J., HELMIG, R. and SCHLÜTER, M. (2002), Numerical simulation of freshwater, salt water and methane interaction processes in a coastal aquifer, *in* J. Chadam, A. Cunningham, R. E. Ewing, P. Ortoleva and M. Wheeler, eds, 'IMA Volumes in Mathematics and its Applications', Vol. 131: Confinement and Remediation of Environmental Hazards and Resource Recovery, Springer, New York, pp. 262–283.

HINKELMANN, R., SHETA, H., HELMIG, R., SAUTER, E. J. and SCHLÜTER, M. (2000b), Numerical Simulation of Water-Gas Flow and Transport Processes in Coastal Aquifers, *in* K. Sato and Y. Iwasa, eds, 'Groundwater Updates', Springer, Tokyo, Berlin, New York, pp. 295–300.

HOLZBECHER, E. (1998), *Modeling Density-Driven Flow in Porous Media*, Springer, Berlin, Heidelberg, New York.

HUMPHREY, A. E. (1972), 'Kinetics of biosystems: a review', *Advances in Chemistry Series* (109), 630–650.

Bibliography

HUYAKORN, P. S., WU, Y. S. and PARK, N. S. (1996), 'Multiphase approach to the numerical solution of a sharp interface saltwater intrusion problem', *Water Resources Research* **32**(1), 93–102.

HWANG, D. W., KIM, G. B., LEE, Y. W. and YANG, H. S. (2005), 'Estimating submarine inputs of groundwater and nutrients to a coastal bay using radium isotopes', *Marine Chemistry* **96**(1-2), 61–71.

INTERNATIONAL FORMULATION COMMITTEE (1967), A Formulation of the Thermodynamic Properties of Ordinary Water Substance, Technical report, IFC Secretariat, Düsseldorf, Germany.

JABIR, A. K. (2005), Finite Volume Models for Multiphase Multicomponent Flow through Porous Media, Dissertation, Institut für Wasserbau, Universität Stuttgart, Stuttgart, Germany.

JABLONSKY, A.-D. (2008), Erweiterung und Optimierung eines Ein-Phasen/Mehr-Komponenten-Modells zur Beschreibung dichteinduzierter Strömungs-, Transport- und Reaktionsprozesse im Untergrund, Diplomarbeit, Fachgebiet Wasserwirtschaft und Hydrosystemmodellierung, Technische Universität Berlin, Berlin, Germany.

JAHNKE, R., HEGGIE, D., EMERSON, S. and GRUNDMANIS, V. (1982), 'Pore Waters of the Central Pacific Ocean: Nutrient Results', *Earth and Planetary Science Letters* **61**(2), 233–256.

JOHANNES, R. E. (1980), 'The ecological significance of the submarine discharge of groundwater', *Marine Ecology-Progress Series* **3**(4), 365–373.

JOURABCHI, P., VAN CAPPELLEN, P. and REGNIER, P. (2005), 'Quantitative interpretation of pH distributions in aquatic sediments: a reaction-transport modeling approach', *American Journal of Science* **305**, 919–956.

KARPEN, V., THOMSEN, L. and SUESS, E. (2006), 'Groundwater discharges in the Baltic Sea: survey and quantification using a schlieren technique application', *GEOFLUIDS* **6**(3), 241–250.

KHALILI, A., BASU, A. J., PIETRZYK, U. and JORGENSEN, B. B. (1999), 'Advective transport through permeable sediments: a new numerical and experimental approach', *Acta Mechanica* **132**(1-4), 221–227.

KINZELBACH, W. (1992), *Numerische Methoden zur Modellierung des Transports von Schadstoffen im Grundwasser*, Vol. 21 of *Schriftenreihe gwf Wasser Abwasser*, R. Oldenbourg Verlag.

KITWARE (2008), 'Paraview', Webpage: http://www.paraview.org/. Last Checked January, 2009.

KOLDITZ, O., KAISER, R., HABBAR, D., ROTHER, T. and THORENZ, C. (1999), *ROCKFLOW - Theory and users' manual. Release 3.4*, Institut für Strömungsmechanik und Elektronisches Rechnen im Bauwesen, Universität Hannover, Hannover, Germany.

KOLDITZ, O., RATKE, R., DIERSCH, H.-J. G. and ZIELKE, W. (1998), 'Coupled groundwater flow and transport . 1. Verification of variable density flow and transport models', *Advances In Water Resources* **21**(1), 27–46.

KURTZ, S. (2004), Grundwasseraustrittsstellen im Sahlenburger Watt: Methoden zur Beprobung von Fluiden und Bilanzierung des Grundwasserausstrom, Diplomarbeit, Universität Bremen, Bremen, Germany.

LANG, G. (1990), Zur Schwebstoffdynamik von Trübungszonen in Ästuarien, Bericht Nr. 26, Institut für Strömungsmechanik und Elektron. Rechnen im Bauwesen der Universität Hannover, Hannover, Germany.

LAPOINTE, B., LITTLER, M. and LITTLER, D. (1997), Macroalgal Overgrowth at Fringing Coral Reefs at Discovery Bay, Jamaica: Bottom-Up Versus Top-Down Control, *in* 'Proceedings of the 8th International Coral Reef Symposium', Vol. 1, pp. 927–932.

LAROCHE, J., NUZZI, R., WATERS, R., WYMAN, K., FALKOWSKI, P. G. and WALLACE, D. W. R. (1997), 'Brown tide blooms in long island's coastal waters linked to interannual variability in groundwater flow', *Global Change Biology* **3**(5), 397–410.

LEE, D. R. (1977), 'Device for Measuring Seepage Flux in Lakes and Estuaries', *Limnology and Oceanography* **22**(1), 140–147.

LEGE, T., KOLDITZ, O. and ZIELKE, W. (1996), *Strömungs- und Transportmodellierung, Handbuch zur Erkundung des Untergrundes von Deponien und Altlasten*, Vol. Band 2, Springer, Belrin, Heidelberg, New York.

Bibliography

LEPEINTRE, F. (1992), Bibliotheque BIEF - Note de Principe et Descriptif Informatique, Report HE-43/92-16, Laboratoire National d'Hydraulique, Chatou, France.

LEWIS, J. P. and NEUMANN, U. (2003), 'Performance of Java versus C++', Webpage: http://www.idiom.com/ zilla/Computer/javaCbenchmark.html. Last Updated 2004.

LGN (2000), 'Topo50. Landesvermessung und Geodatenbasisinformation Niedersachsen, Bundesamt für Kartographie und Geodäsie, CD-ROM. Topographische Karte 1:50 000, Topographische Übersichtskarte 1:200 000'.

LINKE, G. (1979), Ergebnisse geologischer Untersuchungen im Küstenbereich südlich Cuxhaven - Ein Beitrag zur Diskussion holozäner Fragen, *in* 'Probleme der Küstenforschung im südlichen Nordseegebiet, Band 13', Niedersächsisches Landesinstitut für Marschen und Wurtenforschung, Wilhemshaven, pp. 39–83.

MÜLLER, T. J. (1999), Determination of salinity, *in* K. GRASSHOFF, K. KREMLING and M. EHRHARDT, eds, 'Methods of Seawater Analysis. Third, Completely Revised and Extended Edition', WILEY-VCH Verlag, Weinheim, New York, Chinchester, Brisbane, Singapore, Toronto, chapter 3, pp. 41–73.

MOORE, W. S. (1996), 'Large groundwater inputs to coastal waters revealed by 226Ra enrichment', *Letters of Nature* **380**, 612–614.

OLDENBURG, C. M. and PRUESS, K. (1995), 'Dispersive Transport Dynamics In A Strongly Coupled Groundwater-Brine Flow System', *Water Resources Research* **31**(2), 289–302.

OLTEAN, C., ACKERER, P. and BUES, M. A. (1994), Solute transport in 3d laboratory model through an homogeneous porous medium - behaviour of dense phase and simulation, *in* Peters, A . and Wittum, G. and Herrling, B. and Meissner, U. and Brebbia, C. A. and Gray, W. G. and Pinder, G. F., ed., 'Computational Methods in Water Resources X, Vols 1 And 2', Vol. 12 of *Water Science and Technology Library*, Kluwer Academic Publ, Po Box 17, 3300 AA Dordrecht, Netherlands, pp. 521–528. 10th International Conference on Computational Methods in Water Resources, Heidelberg, Germany, Jul, 1994.

OSWALD, S., SCHWARZ, C. and KINZELBACH, W. (1996), Benchmarking in numerical modelling of density driven flow, *in* 'roc. of 14th Salt Water Intrusion Meeting SWIM 96', number Report No 87, Geological Survey of Sweden, Uppsala, Sweden, pp. 32–40.

PATANKAR, S. V. (1980), *Numerical Heat Transfer and Fluid Flow*, Series in computational methods in mechanics and thermal sciences, Taylor and Francis.

PINDER, G. F. and COOPER, H. H. (1970), 'A numerical technique for calculating transient position of saltwater front', *Water Resources Research* **6**(3), 875–882.

PRESS, W. H., TEUKOLSKY, S. A., VETTERLING, W. T. and FLANNERY, B. P. (2007), *Numerical Recipes - The Art of Scientific Computing*, 3rd edn, Cambridge University Press.

REID, R. C., PRAUSNITZ, J. M. and POLING, B. E. (1987), *The Properties of Gases and Liquids*, MacGraw-Hill.

RIEDL, R. and MACHAN, R. (1972), 'Hydrodynamic patterns in lotic intertidal sands and their bioclimatic implications', *Marine Biology* **13**, 173–184.

RODEMANN, H., BROST, E., SCHÜNEMANN, J., NOELL, U., SIEMON, B. and BINOT, F. (2005), 'Gleichstromelektronische Untersuchungen eines mit aeroelektromagnetischen Messungen kartierten Süßwasservorkommens im Sahlenburger Watt unter Berücksichtigung von Äquivalenzfällen und 2D/3D-Modellrechnungen', *Zeitschrift für Angewandte Geologie* **1**, 45–53.

SAFFMAN, P. G. (1959), 'A theory of dispersion in a porous medium', *Journal of Fluid Mechanics* **6**(3), 321–349.

SAFFMAN, P. G. (1960), 'Dispersion due to molecular diffusion and macroscopic mixing in flow through a network of capillaries', *Journal of Fluid Mechanics* **7**(2), 194–208.

SAUTER, E. J. (2001), Executive Final Summary Report of the EU-Project SUB-GATE, Environment & Climate Research Programme (1994-1998), Contract ENV4-CT97-0631.

SAUTER, E. J. and SCHLÜTER, M. (1999), Sub-GATE, Summary Progress Report (End of year 1), Report, Kiel, Germany.

SAUTER, E. J. and SCHLÜTER, M. (2000), Sub-GATE, Summary Progress Report (End of year 2), Report, Kiel, Germany.

SCHANKAT, M., HINKELMANN, R. and SCHLÜTER, M. (2007), Numerical Modelling of Density-Driven Flow and Transport Processes in a Coastal Aquifer - Submarine Groundwater Discharge from Seeps, *in* 'Proceedings of the 32nd Congress of

IAHR, Harmonizing the Demands of Art and Nature in Hydraulics, July 1-6 2007, Venice, Italy', IAHR.

SCHANKAT, M., HINKELMANN, R. and SCHLÜTER, M. (2008a), Development of an Object-oriented Numerical Model for Multi-Component Density-Driven Flow, Transport and Biogeochemical Reaction Processes in the Subsurface, *in* 'Proceedings of the 8th International Conference on HydroScience and Engineering, Sept. 8th-12th, Nagoya, Japan'.

SCHANKAT, M., HINKELMANN, R. and SCHLÜTER, M. (2008b), Modellentwicklung für dichteabhängige Strömungs-, Transport- und Reaktionsprozesse in porösen Medien, *in* 'Forum Umwelttechnik und Wasserbau Band 2, 10. JuWi-Treffen Beiträge zum Treffen junger Wissenschaftlerinnen und Wissenschaftler deutschsprachiger Wasserbauinstiute', Innsbruck University Press.

SCHANKAT, M., HINKELMANN, R. and SCHLÜTER, M. (2009a), DiaTrans - A new numerical model to simulate density-dependent flow, transport and reaction processes in subsurface sediments interacting with seawater, *in* 'Proceedings of the 2nd International Multidisciplinary Conference on Hydrology and Ecology (HydroEco) - Ecosystems interfacing with Groundwater and Surface Water, Vienna, Austria, 20-23 April 2009', University of Natural Resources and Applied Life Sciences, Vienna.

SCHANKAT, M., HINKELMANN, R. and SCHLÜTER, M. (2009b), 'Numerical Modeling of the Tidal Influence on Submarine Groundwater Discharge Rates and the Composition of Pore Waters', *Hydrological Sciences Journal* (**in review**).

SCHARF, F. (2008), Der Grundwasseraustritt im Sahlenburger Watt bei Cuxhaven: Datenerhebung und GIS-basierte Untersuchung der Ursachen udn Charakteristiken, Diplomarbeit, Universität Bremen.

SCHEIDEGGER, A. (1961), 'General theory of dispersion in porous media', *Journal of Geophysical Research* **66**(10), 3273–3278.

SCHEUERMANN, A., VARDOULAKIS, I., PAPANASTASIOU, P. and STAVROPOULOU, M. (2001), A sand erosion problem in axial flow conditions on the example of contact erosion due to horizontal groundwater flow, *in* W. Ehlers, ed., 'Iutam Symposium on Theoretical and Numerical Methods in Continuum Mechanics of Porous Materials', Vol. 87 of *Solid Mechanics and its Applications*, Springer, PO Box 17, 3300 AA Dordrecht, Netherlands, pp. 169–

175. IUTAM Symposium on Theoretical and Numerical Methods in Continuum Mechanics of Porous Materials, Stuttgart, Germany, Sep 05-10, 1999.

SCHLÜTER, M. (2002), Fluid Flow in Continental Margin Sediments, *in* G. Wefer, D. Billett, D. Hebbeln, B. Jorgensen, M. Schlüter and T. van Wering, eds, 'Ocean Margin Systems', Springer-Verlag, Berlin, Heidelberg, pp. 205–217.

SCHLÜTER, M., SAUTER, E., HANSEN, H. P. and SUESS, E. (2000), 'Seasonal variations of bioirrigation in coastal sediments: Modelling of field data', *Geochimica Et Cosmochimica Acta* **64**(5), 821–834.

SCHLÜTER, M., SAUTER, E. J., ANDERSEN, C. E., DAHLGAARD, H. and DANDO, P. R. (2004), 'Spatial distribution and budget for submarine groundwater discharge in Eckernförde Bay (Western Baltic Sea)', *Limnology and Oceanography* **49**(1), 157–167.

SCHULZ, H. D. (2000), Quantification of early diagenesis: Dissolved constituents in marine pore water, *in* H. D. Schulz and M. Zabel, eds, 'Marine Geochemistry', Springer, Berlin, Heidelberg, New York.

SEEBERG-ELVERFELDT, J., SCHLÜTER, M., FESEKER, T. and KÖLLING, M. (2005), 'Rhizon sampling of pore waters near the sediment/water interface of aquatic systems', *Limnology and Oceanography: Methods* **3**, 361–371.

SEGOL, G. (1994), *Classic groundwater simulations - Proving and improving numerical models*, PTR Prentice Hall, Englewood Cliffs.

SEGOL, G., PINDER, G. F. and GRAY, W. G. (1975), 'Galerkin-finite element technique for calculating transient position of saltwater front', *Water Resources Research* **11**(2), 343–347.

SIEMON, B. and BINOT, F. (2001), Aerogeophysikalische Erkundung von Salzwasserintrusionen und Küstenaquiferen im Gebiet Bremerhaven-Cuxhaven - Verifizierung der AEM-Ergebnisse, *in* A. HÖRDT and J. STOLL, eds, 'Protokoll über das 19. Kolloquium "Elektromagnetische Tiefenforschung"', pp. 319–328.

SINGH NOTAY, K. V. (2007), Development of a Density-Driven Subsurface Flow and Transport Model using a Multi-Component Formulation in an Object-Oriented Framework, Masters Thesis, Chair of Water Resources Management and Modeling of Hydrosystems, Technische Universität Berlin; in co-operation with EUROAQUAE, Universität Cottbus, Cottbus, Germany.

STEPHAN, K. and MAYINGER, F. (1990), *Thermodynamik: Grundlagen und Technische Anwendungen*, Vol. Band 1: Einstoffsysteme, Springer Verlag.

TANIGUCHI, M., BURNETT, W. C., CABLE, J. E. and TURNER, J. V. (2002), 'Investigation of submarine groundwater discharge', *Hydrological Processes* **16**(11), 2115–2129.

TANIGUCHI, M., ISHITOBI, T. and SHIMADA, J. (2006), 'Dynamics of submarine groundwater discharge and freshwater-seawater interface', *Journal of Geophysical Research* **111**(C1), 1–9.

TROMP, T. K., VAN CAPPELLEN, P. and KEY, R. M. (1995), 'A global model for the early diagenesis of organic carbon and organic phosphorus in marine sediments', *Geochimica et Cosmochimica Acta* **59**(7), 1259–1284.

VALIELA, I., COSTA, J., FOREMAN, K., TEAL, J. M., HOWES, B. and AUBREY, D. (1990), 'Transport of groundwater-borne nutrients from watersheds and their effects on coastal waters', *Biogeochemistry* **10**(3), 177–197.

VAN CAPPELLEN, P., GAILLARD, J.-F. and RABUOILLE, C. (1993), Biogeochemical transformations in sediments: kinetic models of early diagenesis, *in* R. Wollast, F. T. Mackenzie and L. Chou, eds, 'Interactions of C, N, P and S Biogeochemical Cycles and Global Change', Springer, New York, NY.

VAN CAPPELLEN, P. and WANG, Y. F. (1995), Metal cycling in in surface sediments: Modelling the the interplay between transport and reaction, *in* H. Allen, ed., 'Metal Contaminated Aquatic Systems', Ann Arbor Press, pp. 21–64.

VAN CAPPELLEN, P. and WANG, Y. F. (1996), 'Cycling of iron and manganese in surface sediments: a general theory for the coupled transport and reaction of carbon, oxygen, nitrogen, sulfur, iron, and manganese', *American Journal of Science* **296**(3), 197–243.

VOSS, C. I. and SOUZA, W. R. (1987), 'Variable density flow and solute transport simulation of regional aquifers containing a narrow fresh-water-saltwater transition zone', *Water Resources Research* **23**(10), 1851–1866.

WANG, Y. F. and VAN CAPPELLEN, P. (1996), 'A multicomponent reactive transport model of early diagenesis: Application to redox cycling in coastal marine sediments', *Geochimica Et Cosmochimica Acta* **60**(16), 2993–3014.

WHITICAR, M. (2002), 'Diagenetic relationship of methanogenesis, nutrients, acoustic turbidity, pockmarks and freshwater seepages in Eckerförde Bay', *Marien Geology* **182**, 29–53.

ZHENG, C. (1990), A modular three-dimensional transport model for simulation of advection, dispersion, and chemical reactions of contaminants in groundwater systems, Report to the Kerr Environmental Research Laboratory, US Environmental Protection Agency, Ada, OK.

ZIPPERLE, A. and REISE, K. (2005), 'Freshwater springs on intertidal sand flats cause a switch in dominance among polychaete worms', *Journal of Sea Research* **54**(2), 143–150.